MW01530737

Masculine, Feminine, and Fully Human

Developmental Paths Through the Adult Years

Richard W. Coan

authorHOUSE®

AuthorHouse™
1663 Liberty Drive, Suite 200
Bloomington, IN 47403
www.authorhouse.com
Phone: 1-800-839-8640

© 2008 Richard W. Coan. All rights reserved.

No part of this book may be reproduced, stored in a retrieval system, or transmitted by any means without the written permission of the author.

First published by AuthorHouse 11/14/2008

ISBN: 978-1-4389-2173-0 (sc)

Printed in the United States of America
Bloomington, Indiana

This book is printed on acid-free paper.

CONTENTS

1. Introduction

For thousands of years, the differences between boys and girls and between men and women have been a frequent topic of comment and discussion. The issue has been pursued in recent times by philosophers, psychologists, theologians, and popular writers. Many questions recur over and over as people pursue this topic. Just what are the differences between the sexes? Is there an unbridgeable gap? Do we just imagine differences where none exist? Is one sex better than the other? Should we seek to enhance the differences or diminish them? Are there potentials men and women should nurture to enhance their personal growth through the adult years?

The comparison of males and females has often served to promote the view that the members of one sex are superior to those of the other. Thus the playful verses of Mother Goose highlight the flaws of boys:

> What are little boys made of?
> What are little boys made of?
> Frogs and snails
> And puppy-dog tails,
> That's what little boys are made of.

Girls on the other hand are said to be made of "sugar and spice and all that is nice." Subsequent verses inform us that young men are made of "sighs and leers and crocodile tears," while young women are composed of "ribbons and laces and sweet pretty faces."

In the view of Mother Goose, girls and young women pose few problems so long as they conform to their standard image. The description of little boys, however, accords with the common impression that boys tend to be messier and more prone to behavioral problems than girls. The subsequent verse indicates that young men display greater outward emotionality than do young women, but do they in fact or is this quality subject to the fashions of the time and place? Men are more often accused of being emotionally insensitive, and one does not expect the eruption of crocodile tears when the surrounding culture dictates that "big boys don't cry."

The flaw most often ascribed to the male members of our species is a proneness to violence and aggression. Nonetheless, given the patriarchal ethos that has long prevailed throughout the Middle East and the Western world, women have more commonly been depicted as inferior to men, particularly with respect to intelligence. Women have been said to have the minds of children, and even well into the twentieth century, a few men in prominent positions have declared that women lack the intellectual gifts they would need to excel in the more demanding fields of science. Needless to say, the intellectual and artistic attainments of women have always fallen short in times and places where women have been denied the educational opportunities available to men. In recent times in the Western world, societal pressures and expectations have been weighted in favor of men for most professional fields.

In theological writings, women have sometimes been characterized as morally weaker than men, and the myth of Adam and Eve is cited as evidence. It is Eve who first succumbs to the temptation to eat the forbidden fruit of the tree that stands in the middle of the garden of Eden, and she then gives some of the fruit to Adam. Since this is the fruit that provides the knowledge of good and evil, we can construe the story as evidence either that the woman is more prone to sin than the man or that she is more open to moral awareness. Of course, the former interpretation has tended to prevail in theological discourse. Nonetheless, it was the action of Eve that initiated the emergence of our mythic forebears from the infantile realm of witless bliss.

In early Christianity two female images played a prominent role, that of the Virgin Mary and that of Mary Magdalene. In the former case, we have a woman said to be so devoted to God that she bears the son of God and, though married to Joseph, remains a virgin throughout her life. An early pope declared that Mary Magdalene on the other hand was a fallen woman who was redeemed by her devotion to Christ. He thus equated her with another woman named Mary whose story appears in the scriptures, and by doing so he borrowed a mythic theme that predated Christianity. The images associated with both the Virgin Mary and Mary Magdalene assume that sex per se is sinful, particularly when indulged in by women outside the sacred bonds of matrimony. A woman can be totally pure and virtuous only through complete abstention from sex.

The two images are only roughly supported by the texts of the four gospels adopted by the Western church in the fourth century, and there is no historical evidence outside those texts to support them. It is quite unlikely that the mother of Jesus was in reality a perpetual virgin. The image of Mary Magdalene is equally fanciful. In some of the gnostic gospels suppressed by the men in power in the early church, we are told that she was the disciple closest to Jesus and the one who best understood his mission. And more than one historian has argued that she was in all likelihood the wife of Jesus.

Over the years, men and women have been contrasted with respect to many qualities in addition to their respective mental and moral gifts. Most of us find it difficult to view people simply as people. Their identity as members of one sex or the other seems to be important. I remember sitting in a large lecture hall some years ago when a blond student across the room caught my attention. Given the short hair style and unisex clothing, I could not immediately tell whether I was seeing a young man or a young women, and I felt a need to know. I soon decided on the basis of her movement patterns that I was looking at a young woman. Then I was able to let the matter rest and direct my attention elsewhere.

Because the perception of the individual's sexual status is so significant, we are fascinated by male and female impersonators who are adept at convincing us that they are not what they seem. It is interesting to watch movies like *Tootsie* and *Her Life as a Man* and marvel at the successful

3

deception. This was true in Shakespeare's time, for in some of his plays, women disguise themselves and pretend to be men. No doubt the situation would have been a source of amusement for members of the cast, since it was common on the stage in the Elizabethan age for the roles of women to be assumed by male actors, suitably attired for the occasion. Thus, there were male actors playing the roles of women who pretended to be men.

Because the biological status of the person with whom we interact seems so significant, we are inclined to look for distinguishing features beyond the essential anatomical ones. For this reason, people have long talked about the basic differences between men and women. It is inevitable that much of what they have perceived is either a misinterpretation of reality or an exaggeration of slight differences between overlapping distributions. Whatever differences are alleged to exist, whether real or imaginary, it is possible to argue that either sex is somehow superior to the other. I believe it is reasonable to say that such a conclusion serves primarily to highlight the value bias of the individual making the assertion.

In recent years two positions regarding sex differences have received considerable attention. One is the view that there are really no inherent psychological differences between men and women. This implies that many of the differences we perceive are more apparent than real. We merely perceive what we expect to perceive. Of course, there are some actual differences in personal qualities and intellectual performance that we can identify, but these are regarded as differences created by our society as a result of the different treatments accorded to boys and girls from birth on into the adult years. This is a position advanced by a few feminists in the course of their worthy efforts to bring about an end to various forms of discrimination against women.

The other position is that there are important differences between men and women and that they need to be understood and respected. It is a rare marriage in which the husband and wife are so alike in intellect and temperament that they always see eye and eye and respond alike to every situation that arises. They will nearly always differ in some respects, and it is well if each can value the ways in which the spouse is different. If there is a recurring pattern of differences between husbands and wives, it is well for all of us to recognize these, so we can better appreciate our mates for the individuals they are, rather than demanding that they shape up and become more like ourselves. If there are consistent differences between men and women, it behooves our society to recognize the value of the contributions more likely to come from either sex. Otherwise, we may simply complain about the thick skulls of men, or like Henry Higgins in *My Fair Lady*, we may cry, "Why can't a woman be more like a man?"

In this book we shall consider what is known about sex differences. These include differences observed in infants, differences found during the years of childhood and adolescence, and differences evident in adult men and women. Of course, none of us passes through the decades of youth, middle age, and the later years of life without undergoing additional changes in our awareness and behavior. There are some common patterns to these changes through the adult years, but no one pattern appears to be universal. As we note these changes in our friends and relatives, some of them appear to be improvements, while others seem regrettable. Is there one ideal developmental path that we should all seek to follow?

This is a question that has been answered in more than one way. What are the qualities that men and women need to realize and nurture during their adult years. There are writers who believe that most men in our society have failed to realize fully the qualities that are most essential to the masculine psyche. To varying degrees they have learned to suppress those qualities and play the roles prescribed and demanded by our society. They have paid a price in order to gain social acceptance. For full development, men must get in touch with their true nature.

They are other writers who have offered a similar message with respect to women. They note an assortment of feminine potentials embodied in ancient mythic images that women in our society have failed to realize. Perhaps they would argue that women have paid an even heavier price than men, for they have been offered a narrower range of acceptable roles. These two sets of writers have been running on parallel tracks, each stressing the need for members of a given sex to realize their most natural qualities. If they were all to do so, would we see greater differences between men and women? This is certainly not a given, for it is possible that the natural differences are not as great as those demanded by society.

There is still a third group of writers who see the ideal condition as one of androgyny, a blending of masculine and feminine qualities. They would contend that both sets of qualities are an inherent part of each of us. Some of these writers are inspired by theorists like Carl Jung, who held that the anima (the unconscious, feminine soul) is an inherent part of the psyche of men, while the animus (the unconscious, masculine soul) is an inherent part of the female psyche. In the view of these writers, we all achieve wholeness to the extent that we expand our awareness of the contrasexual side and embrace both sides of our nature.

The first two positions are sometimes viewed as opposed to the third. Yet if we are interested in the full development of our human potentials, there is certainly merit in each of these positions, and they are not inherently incompatible. I also see drawbacks to pursuing either the first or second position to the exclusion of the third. We do this if we imagine that seeking androgyny means denying what is either most distinctively masculine or most distinctively feminine. Conversely, seeking to realize fully either the most natural masculine or the most natural feminine qualities does not mean that we must forego a realization of the other set of qualities. As I see it, the most essential question is: how can we best realize all our potentials as human beings?

The moment we pose this question, another one begs to be answered: what are the potentials we, as human beings, might hope to realize? Are there possible ways of living, acting, thinking, or being that we have disregarded and need to recognize? Most of us have spent our lives living in this society, and we tend to focus on modes of living characteristic of this social setting in the modern age. Furthermore, we are accustomed to thinking of certain modes as appropriate for men and others as appropriate for women.

Whatever there may or may not be in the way of inherent differences between the sexes, we should expand the scope of our vision and look beyond the local contemporary scene. We can do this by studying many other cultures. A strategy that may prove even more useful, however, is to examine the themes that run through major mythologies.

As a number of other writers have recognized, the significant figures in mythologies — gods, goddesses, heroes — embody various distinctive modes of being. Of course, these figures are shaped by the societies that gave rise to their myths. Yet they represent themes that have endured over many centuries, and to a great extent these themes transcend any given cultural setting. The qualities displayed by a significant figure in one major mythology may resemble those shown by significant figures in other major mythologies.

To the extent that the recurring themes of mythology transcend the boundaries of time and place, we can argue that they represent major modes of being that are an inherent part of our nature as members of the human species. It is tempting to think of some of these modes as masculine and others as feminine. It is true that some modes appear more often in male figures, while others appear more often in female figures. Yet we find exceptions with respect to every mode. We restrict our understanding needlessly if we try to classify all these modes as either masculine or feminine. They are all human modes of being available to each of us.

We may think that our personal lives are simply guided by the reality of our individual constitutions and our surroundings. Yet each of us begins early in life to develop a personal myth. It includes a personal image, a sense of the kind of person we are, and it includes a life story that we write inside our heads as we grow up and pass through our adult years. That personal myth incorporates one or more themes that we can find in major mythologies.

Our own personal myth may serve us well at one stage in life and then prove maladaptive. If we cling to it too closely, we can limit our freedom to grow. It is my hope that by exploring the major themes of mythologies we can all find fresh ways of coping with the issues in our lives and continue developing in rewarding ways through the years that lie ahead.

Are there specific modes of being that we all need to cultivate? Are there some that men should cultivate and others that women should cultivate? Is there one ideal path to human fulfillment that everyone ought to follow? I am not inclined to answer any of these questions with an unconditional *yes*. I believe that many different developmental paths are possible. I also believe that there are several alternative developmental goals toward which we may strive. In this book, I seek to highlight the possible choices.

In this introductory chapter I am raising all these issues in a broad, abstract manner. In the pages that follow, I hope to lay the groundwork that will permit us to consider them in greater depth. Throughout this book, I shall encourage you to reflect on your own life and consider how the ideas we are considering apply to yourself. I urge you to entertain possibilities for exploring new potentials for growth, awareness, and joy in living. I shall also share bits of my own ongoing journey as a fellow traveler through life — and as a fellow seeker who does not claim to have found all the answers as yet.

2. The Basic Differences

How do men and women really differ? It would be virtually impossible to catalog everything that has ever been said about this. Much research has been aimed at identifying differences, and researchers commonly find slight differences between the sexes when they are not looking for them.

A few writers have suggested broad principles that govern what they see as the essential features that distinguish men from women. David Bakan (1966), for example, speaks of two basic modalities that characterize living organisms. These are agency and communion, which may be regarded respectively as masculine and feminine principles. Agency reflects a concern with self and is expressed in self-assertion, self-expansion, and self-protection. Communion implies a concern for others and an urge to seek relationship.

Simon Baron-Cohen (2003) makes a related distinction, contending that the female brain is predominantly hard-wired for empathizing, while the male brain is predominantly hard-wired for systematizing. Empathizing implies something similar to communion, but the emphasis is on the feeling or emotional quality that enables us to grasp what the other person is experiencing so that we are able to connect and provide care. Systematizing, on the other hand, has to do with exploring, analyzing, and constructing systems. It is akin to the concept of agency in implying the separation between self and other that enables us to act in ways different from those implied by the concepts of communion and empathizing.

John Gray (1992) pursues much the same theme in his book *Men Are from Mars, Women Are from Venus*. He contends that men are more interested in things and that they value power, competency, and achievement. Women in contrast are more interested in people and feelings, and they value love, communication, relationships, and beauty. It is easy to read into these descriptions something like communion and empathizing on the female side as opposed to agency and systematizing on the male side. Gray's book is aimed at facilitating communication between husbands and wives (and between unmarried partners), and he points to differences in styles of interaction that everyone needs to recognize. He says that men tend to cope with stress by withdrawing and focusing on problem-solving, while women tend to cope by talking and sharing feelings. He believes that men are motivated when they feel needed and that they want to receive trust, admiration, and encouragement. Women on the other hand have a greater need to feel cherished, and they want to receive devotion and understanding.

I am sure that none of these authors believes that men and women fall neatly into two clearly separated camps. In distinguishing between a masculine principle and a feminine principle, we are merely talking about tendencies that predominate in either men or women. However these principles are expressed in thinking and action, we would expect to find two overlapping distributions. The average man and the average woman might not be very different from each other. Any individual who functions only at the masculine extreme or at the feminine extreme, manifesting none of the other quality, is likely to have serious problems in coping with life, the world, and other people.

The terms *masculine* and *feminine*, of course, can encompass a vast number of features of behavior and experience in which there are differences between the sexes. Now I must note that it has become fashionable in recent years to speak of what everyone used to call *sex* differences as differences in *gender*. The term *sex* in this case would refer to strictly biological features, features of anatomy, physiology, and biochemistry, while *gender* would refer to roles and other attributes deemed appropriate in a given society. The term *gender* is subject to some variation in usage, however, because contemporary writers vary in the extent to which they regard the qualities that it encompasses as resting firmly on a biological basis or as being merely societal impositions.

I believe we must recognize that nearly all the differences we observe in behavior and consciousness between men and women rest on an interplay of biological and social determinants and can rarely be ascribed totally to just the one source or the other. Even where biology plays a powerful role, the masculine or feminine quality need not be immutable. In sum, I am willing to grant that there is some value to extending the meaning of the word *gender* beyond its traditional meaning as a grammatical term, though we are not really clarifying our thinking very much by the expanded usage of this word. In any case, let us proceed to consider what is known about some of the differences between the sexes.

GENDER IDENTITY AND IDENTIFICATION

If by *sexual identity* we mean the biological status of the individual, whether that person possesses the usual male or the usual female reproductive organs, *gender identity* would mean the awareness that one is a boy, girl, man, or woman. Whatever the nature of the individual's gonads and the corresponding internal and external anatomical features, the child is assigned by doctor and parents to one sex or the other at birth, if not before. That label continues as the infant grows, and normally in the second year of life, the boy comes to understand that he is a boy and is thus a member of the category that also includes his father, brother, grandfather, uncle, and any other males that he meets. The girl at the same time comes to know that she is a girl and is thus like her sister, mother, aunt, and grandmother. With rare exceptions, gender identity corresponds to sexual identity.

In early childhood, of course, the child may have only a rudimentary sense of what this identity implies. An awareness of the male-female differences in sexual anatomy is subject to variation from one family or setting to another. In the early years, children may come to know the differences between males and females primarily in terms of clothing, hair styles, and various bits of behavior. In the course of childhood, gender identity expands to encompass more of what one comes to know about the essential nature of each sex.

Gender identity is obviously a key part of the experience of each of us, and it usually remains quite set from an early stage in life. There are some related facets of our experience to consider, and they have to do with our ability to sense the experience of someone of the other sex. The first is the extent to which we can identify or empathize with the other person. A man with a strong masculine gender identity may still be able to identify with a woman who is upset, fearful, joyful, or sad. Correspondingly a woman with a strong feminine gender identity may be able to identify with a man in a strong emotional state. This ability is subject to much individual variation, but it is possible that women on the average are more adept at this than men are.

Then there is the issue of one's characteristic identification. A man may have a clear sense that he is a man yet feel that he is more like the women he knows than he is like the men he knows. Conversely a woman who clearly recognizes that she is a woman may identify more with men than with women. We would expect that, as a rule, basic gender identity and identification would be either both masculine or both feminine, but I suspect that exceptions to the rule are very common.

One other facet is imaginal freedom, the extent to which one is able to imagine really being a member of the other sex. In some of the psychology classes I have taught, I have used experiential exercises that enable students to gain a better understanding of their own personalities and further their own personal development in whatever way they wish. One type of exercise is a fantasy structure. While the students sat with eyes closed, I would read the beginning of a fantasy and then allow them time for the fantasy to continue however it might inside their heads. When the students felt the fantasy had come to completion, they would write what they had experienced.

One fantasy structure took this form: "You have been taking a trip by yourself to a distant part of the country by bus. The bus on which you have been riding has arrived at the end of its line in a city where you have never been before. Late at night you walk a short distance from the bus depot to a hotel where you obtain a room for the night. You are very tired and you retire at once and fall sound asleep. In the morning you awaken in your hotel room. It is light outside and there is quite a bit of light coming through the shade on the window. Then you realize that you are different. You feel your face, your arms, your hands, your chest. You look at your body, and you see that you are now a person of the opposite sex from what you were before you fell asleep."

Another fantasy structure ran as follows: "You are interested in experiencing hypnosis, and an adept and highly regarded hypnotist offers to provide you with a session. You lean back in a comfortable chair and listen closely as he proceeds with the induction. You are soon in a deep trance with your eyes closed. He guides you back to earlier times in your life. You soon experience yourself as a ten-year-old once again. You feel what it's like to be back in the body of a child, and you think of your friends and your classroom. Then you go back to age 5 and then age 2. You are a lot smaller at this age. But then the hypnotist takes you back to a time several years before your birth. You look around and notice that your surroundings look familiar. Yet you are now in a much different body and leading an earlier life. Who are you and what sort of things do you do?"

I used a variety of other fantasy structures, but I never used more than one on any given occasion. The two I have noted above obviously have something in common, and the student who experienced great imaginal freedom in one was likely to experience great imaginal freedom in the other. Waking up to find that one's sexual identity is transformed is an opportunity to experience a much different body from the inside. This can be terrifying, depressing, or fun when one remains in the hotel room, and stepping outside to deal with other people in that setting presents a new challenge. It is my impression that women students on the whole found this a little easier than did the men, but there was much individual variation.

In completing the other fantasy, most students imagined themselves as having the same sexual identity in the earlier lifetime that they had in the present, but there were exceptions. It was rare for a male student to imagine an earlier life in which he was a woman, but more common for a female student to imagine an earlier life as a man. I suspect we would find the same pattern if we surveyed all the people outside my classes who actually claim to remember previous lives. This is all consistent with the impression that women in general have a little more freedom than men to identify with the other sex and to sense what it is like to be a member of the other sex. It is possible that men are more likely to feel threatened by the prospect of sensing what it is like to be a woman. The fact that we live in a society with patriarchal roots that has long considered men more important than women may have something to do with this. Unfortunately I can offer no data on societies with the opposite bias.

SEXUAL ORIENTATION

Sexual orientation, the sexual attraction to people of the same or the other sex, is another obvious area of sex differences. Since the great majority of people are heterosexual, one may consider attraction to females as a masculine trait and attraction to males as feminine. These traits are correlated to some extent with other masculine and feminine traits, but the association is not inevitable. A male homosexual may be quite masculine in other respects, and a lesbian woman may be generally quite feminine.

We should also note that people do not all fall neatly into the categories of heterosexual and homosexual. There is a continuum of gradations between the two. Alfred Kinsey and his associates used a seven-point rating scale in their research, a rating of 0 meaning exclusively heterosexual and a rating of 7 meaning exclusively homosexual.

An individual may shift along this scale over the years. It is hard to say at what point in life anyone can first be clearly typed with respect to sexual orientation. Undoubtedly many children by age 5 or 6 are aware that they find the genitals of other children, most often those of the other sex, to be particularly interesting, perhaps even exciting. The interest grows and tends to increase markedly with the onset of puberty.

Heterosexuality is the orientation prescribed by our society. For this reason, many people who do not feel strongly drawn to the other sex will still abide by the standard rules and seek a partner of the other sex. They may decide after some years of marriage that this arrangement really does not suit their natural dispositions. Their marriage come to an end, and they then seek partners of their own sex.

Some bisexual individuals, of course, manage to move back and forth over the years between male and female partners. Most surveys indicate that exclusive homosexuality is a little more common in the male population than in the female population. There is reason to believe, however, that there are somewhat more women than men who are able to regard the sexual identity of the partner as relatively unimportant. Thus, they may be able to move in a more fluid way between partners, feeling love for the partner as a person, rather than as someone who must be either a man or a woman.

The fact that some individuals shift from a heterosexual to a homosexual lifestyle at some point in their adult years does not necessarily mean that their underlying disposition has undergone a radical change. It is more often the case that the homoerotic urge was present earlier but was suppressed. Feeling pressure to conform to the ways of the cultural milieu, the individual may have decided the urge was inappropriate or that it would wither in time and give way to the pleasures that marital sex would provide.

In some cases, the individual finds the homoerotic urge so unacceptable that its very existence is denied. For this reason, I believe we need to make a distinction between the inherent sexual orientation of the individual and the explicit sexual orientation. In most people there may be little difference between the two. There is a marked discrepancy, however, in the case of the individual with an inherent homosexual disposition who finds he or she must not only repress that urge but replace it with a heterosexual urge.

We see this most clearly in the homophobic male. Homophobia is most often a male problem in our society. Many men sense any homoerotic urge as a serious threat to their status as real men. When confronted by other men who are overtly homosexual or who act as if they might be — or, even worse, men who make seductive advances toward them — they may become quite agitated and resort to violence. At other times, they may engage in vigorous sexual exploits with women in an effort to prove that their basic orientation is really heterosexual.

INTELLECTUAL ABILITIES AND COGNITIVE STYLE

Many studies have sought to assess differences in overall intellectual performance, but I shall not attempt an extensive survey of the evidence. Many other writers in past years, notably Eleanor Maccoby and Carol Jacklin (Maccoby and Jacklin, 1974), have already done an admirable job of compiling it. The overall verdict is clear. Men vary widely in intelligence, and so do women, but despite many dogmatic pronouncements to the contrary, there is little evidence of a basic difference between the sexes. In a given study the average score for boys or men may be a little higher than that for women or girls, but it is just as likely to be a little lower.

In any given study, the difference in averages is usually too small to reach statistical significance. The difference that appears also varies according to the content of the test, for some kinds of items tend to favor males while others favor female subjects. It is at least partly for this reason that the intelligence tests most widely administered to individual subjects — the Stanford-Binet and the Wechsler tests — contain a balance of different kinds of items, with the result that there is little or no difference in overall performance between male and female subjects at any given age level.

Numerous studies have focused on the verbal abilities of children. Whenever a difference between boys and girls has been found, it has always been small but has most often favored girls. Girls appear to have an advantage with respect to verbal fluency, while it is clear that boys are more likely to exhibit speech and language problems. Both stuttering and dyslexia, for example, are far more common in boys than in girls.

When boys and girls have been compared on tests of quantitative ability, the results have usually favored boys, but a comparison of earlier and more recent studies suggests that the male advantage is declining. Boys have most often been found to perform better than girls on some tests of visual-spatial ability, notably those that involve mental rotation of visual figures. There are many forms of visual-spatial ability, however, and for many of them, little or no difference has been found between boys and girls.

It is widely believed that men are more likely to possess a high level of creativity than women for the obvious reason that the vast majority of people recognized as creative geniuses in the arts and sciences have been men. The evidence from studies comparing boys and girls or men and women, however, has not been so clear because creative ability defies simple measurement. The tests employed have usually involved some kind of ideational fluency, such as the ability rapidly to generate hypotheses or to suggest unusual uses for common objects. Such tests are often as much measures of verbal fluency as they are of creative thinking. The challenge faced by the test constructor is that the most creative subject may respond to the test in ways that the test constructor is unable to anticipate.

Studies of highly creative people indicate that they often lead very unconventional lives. They are often willing to do novel things in various aspects of their lives because they feel free to be different, to be nonconforming to the prevailing modes. I believe it is fair to say that such a lifestyle is more common for men than for women, but the reason is not that clear. One might argue that girls and women have a stronger natural inclination to seek smooth relationships and acceptance, while boys and men can more easily act on impulses without regard for the social consequences. At the same time, however, it appears that our society demands more conformity on the part of its female members, while boys and men are accorded more freedom to be deviant and to "do their own thing."

The most influential musical composition of the twentieth century was Igor Stavinsky's *Rite of Spring*, because it set the stage for other composers to do novel experiments with melody, rhythm, and harmony. It is hard to imagine a woman composer daring to produce such a work, for it evoked a near-riot among the people who had come to attend its first performance in 1913. The composer himself may have felt a bit stung by the complaints and harsh words that ensued, for he showed more restraint in his compositions during the next few years.

The kind of difference between men and women that I have suggested, favoring greater creativity on the part of men, is consistent with the broad principles proposed by the writers I noted at the beginning of this chapter. The individual willing to be different and think independently would manifest more of the principle that Bakan calls agency and, as a "Martian" male, would attach greater value to power, competency, and achievement. The individual more intent on seeking relationships, sharing feelings, and empathizing would be more concerned with understanding the thoughts of others. It is possible that the differences between men and women are less a matter of a difference in ability than a matter of a difference in cognitive style - the style of thinking, perceiving, and knowing. The masculine style may favor creativity, while the feminine style may favor empathic and intuitive understanding. It is even possible, as James Stephens noted, that "women are wiser than men because they know less and understand more."

SOCIAL BEHAVIOR

Many studies have focused on the social behavior of male and female subjects. The clearest, most consistent finding is that physical aggression in more evident in males at all stages of childhood. Furthermore, far more men than women are arrested and convicted for violent crimes.

From infancy on we can observe more rough play among boys than among girls. In the course of this play, boys may collide with one another, but the intention is not necessarily aggression. This result may simply be a function of the higher activity level found among boys. A related finding is that boys are more likely to appear power-oriented and to seek dominance over their peers.

It might be well, however, to qualify all these findings by noting that boys tend not only to be more active than girls but also bigger and stronger. The physical advantage of males is still more marked at the adult level. For this reason alone, it is more risky for a female to be physically aggressive, at least in interaction with a male. If a girl or woman has an aggressive urge, she may find a mode of expression less direct than an overt physical act. She may resort to verbal aggression or to passive non-compliance, or she may seek vengeance toward a disliked peer by talking to others in her peer group. A girl or woman may also manage to get her own way in interacting with boys or men by tactics more subtle than an overt assertion of dominance. It is not uncommon to find a married couple in which the husband is certain that he is the dominant head of the household, even though all the major decisions are the ones chosen by his wife.

Our conventional notions of femininity lead us to expect more in the way of positive social interaction and a greater concern for the feelings of others on the part of girls and women. Pertinent studies have been conducted over the course of many decades. Thus, Lewis Terman and Catherine Cox Miles (1936) reported that females were more compassionate, sympathetic, and emotionally expressive than males. Their evidence came largely from self-reports. In a later research review, Maccoby and Jacklin (1974) found that in studies where behavior was actually observed there was little evidence that female subjects displayed more positive social interaction than males did.

In a still more recent book, Hillary M. Lips (2005) concluded that female subjects are more likely than males to display such positive social traits as nurturance, empathy, and altruism when the evidence consists of self-reports. It appears that these traits are also more likely to be observed in female subjects in situations where they are aware that they are being observed, and Lips suggests that the behavior found in such research may reflect an effort on the part of many female subjects to play the roles for which they have been strongly socialized. What we observe in a laboratory setting, of course, may not be typical of the behavior that characterizes people in everyday life, but that behavior is much more difficult to sample.

INTERESTS, PREFERENCES, AND SELF-DESCRIPTION

In 1936, Terman and Miles reported that men on the average expressed greater interest in adventure, outdoor and strenuous activities, machinery and tools, science, physical phenomena, inventions, and business and commerce. In contrast, women on the average expressed more interest in domestic affairs, aesthetic objects and occupations, sedentary and indoor occupations, and occupations that involve helping the young, helpless, or distressed. These results are consistent with findings obtained in subsequent years with other interest inventories.

A related sex difference has been found consistently by those who have used the six scales of the Allport-Vernon Study of Values. Men in general score higher than women on theoretical, economic, and political values, which would suggest a greater interest in abstract ideas, practical success, and power. Women tend to score higher on aesthetic, social, and religious values, which would indicate a greater interest in the arts, in religion, and in the welfare of others.

Findings that have been reported for various personality inventories are generally consistent with the differences we have noted for social behavior. Furthermore, the responses of women and adolescent girls generally point to greater sensitivity, more neurotic tendencies, greater introversion, less self-confidence, and less self-sufficiency than found for male subjects. Before we jump to broad conclusions from this about feminine personality traits, however, we must recognize that women on the whole are more willing than men to report weaknesses and difficulties. This is one reason they seek medical help more readily than men do — and, perhaps, why they the tend to live a few years longer than men. We must also note that the responses obtained on self-report inventories just tell us something about the way in which people view and describe themselves, but self-description need not correspond in a simple one-to-one way to behavior. People may not be as virtuous, as nasty, or as inadequate as they say they are.

We shall consider some additional findings for the self-descriptions of men and women in the next chapter. But now, we might consider some findings obtained in test situations involving something a little different from self-description. Some years ago, working in collaboration with Raymond B. Cattell (Cattell and Coan, 1959), I conducted an extensive study of the personalities of children in the primary grades. In addition to questionnaire items, I devised a variety of test situations requiring the children to respond to items of many other kinds. One was a test of sound preferences. Sounds were presented in pairs, and the child was asked to indicate a preference for one of the sounds in each pair. Each pair contained one sound that was relatively commonplace and peaceful and one sound that was more unusual and possibly disturbing. For example, I had the sound of a woman singing paired with the sound of a woman screaming. And there was the sound of a man laughing benignly paired with the man laughing diabolically. (I must confess that I enjoyed producing and recording that pair by myself one evening when there was no one else present in the little building where I worked.)

Another of my tests involved picture preferences. Pictures were presented in pairs, and each pair contained a picture of something fairly ordinary and a picture of similar content that was more unusual or strange. For example, I had a sketch of an automobile much like one that the children would have seen on the street paired with a sketch of an automobile with a rather grotesque shape. Gender differences were not the primary focus of my research, but it was quite clear in these two test situations that boys were much more likely than girls to select the item, sound or picture that was more unusual, strange, or disturbing. This strikes me as consistent with other behavioral evidence indicating a greater tendency on the part of boys for impulsive acting-out, non-conformity, and a willingness to be different.

In the course of my teaching career, one of my graduate students, Ruth E. Fehr (1963), conducted a study of color harmony for her master's thesis. She presented many pairs of color patches to two different samples of college-student subjects. The subjects in the first sample were asked to indicate for each pair whether they thought the two colors blended smoothly and harmonized or whether they appeared to clash, creating a more jarring effect in combination. In the second sample, the subjects were shown six pairs of colors at a time and asked to rank the six pairs according to how much they liked each combination.

The researcher found no significant difference between the men and the women in the first sample. They agreed basically with respect to the combinations they considered harmonious and those they considered disharmonious. In the second sample, she found a much greater tendency for the men to show a preference for the color combinations that had been judged as jarring or disharmonious. This findings appears consistent with an overall gender difference involving a greater inclination on the part of boys and men to seek adventure, to take risks, and to act in non-conforming ways.

Erik H. Erikson (1964) devised another test situation that yielded a striking difference between male and female subjects. He saw 300 preadolescent children one at a time, giving them a task that required them to construct a scene with toys on a table. The toys included a varied assortment of human figures, various animals, automobiles, pieces of furniture, and a variety of blocks. Each child was asked to create a scene for a movie, then tell the plot for the scene. The girls tended to produce pleasant scenes inside houses. The boys were more likely to produce high walls and towers and outside scenes involving vehicles and animals, and their plots more often involved collapsing structures and ruins.

Erikson suggests various interpretations of his results but leans toward a psychoanalytic view in terms of projections of the "groundplan of the human body," with girls emphasizing interior space and the boys exterior space. I would note, however, that what is most consistent with what we have found in other test situations is that the girls' scenes tend to be peaceful and harmonious, while the boys' scenes were more likely to involved risk, danger, and strong emotion.

EXPRESSIVE MOVEMENT

In the preceding chapter, I said that I was able to identify a particular student at a distance as a young woman on the basis of her movement pattern. It is possible that I was mistaken, but it is nonetheless true that when we grow up in this country we come to expect certain differences in expressive movement between men and women, even if we pay little or no conscious attention to the differences. Men are more likely to make movements with the arms extended, while women tend to keep their arms close to the body and hold their hands up closer to the chest or face. They make more expressive use of their hands than men do. Women's faces tend to appear more mobile, with varying expression, while men's faces seem more often devoid of expression. As they walk or change position, women more often allow their hips to sway from side to side. On average, women speak more rapidly than men, and their voices display more pitch modulation.

The differences in movement tend to be minimized in the athletic realm. Track, field, and gymnastic feats require essentially the same uses of the body whether the athlete is a man or a woman. The same is true of the arts that involve body movement. Thus, in classical ballet, modern dance, and mime, we usually see little difference between the movements of male and female performers. There is an obvious exception in the case of a performer who is seeking to depict a particular type of person. Then we may see a parody of a stereotypic masculine or feminine movement pattern.

Anyone seeking to impersonate someone of the other sex will try to adopt more of the stereotypic pattern. The man who would be a female impersonator will use gestures he might not otherwise employ, while the woman impersonating a man will exercise more than usual restraint. It is a bit easier for the male performer to achieve a comic effect, if that is the intention, but the female performer can accomplish this with the use of very heavy and clumsy gestures and gait. Away from the stage, the impersonator we are most likely to encounter would be a drag queen posing as a woman. To the extent that he is seeking attention, he is likely to exaggerate the stereotypic feminine patterns.

The gender differences I have just described are typical of American society and to a lesser extent European society, but there is regional and ethnic variation even within the United States. In European societies we see much more use of hand gestures around the Mediterranean than we do in countries to the north. When we proceed to observe people in Asian, African, and Polynesian societies, we may well conclude that few of the gender differences with which we are familiar are at all universal.

3. A Bit of Self-Assessment

Dear reader, much of what I have to say in this chapter will appear more meaningful if you can see how it applies to you personally. So before I launch into an explanation, I suggest you take a few minutes to respond to the following items.

If you wish to avoid marking in the book, just take a sheet of paper and write the item numbers from 1 to 40 down the page. You may want to write them in two or more columns. Leave just enough room to the right of each item number to record a numerical response.

The items below contain words or phrases that have been used to describe people. For each item, you are to decide how accurately the word or phrase describes you. Then record your decision by writing a number from 1 to 4.

1 =not true
2 =slightly true
3 =mostly true
4 =definitely true

1. active
2. bossy
3. committed

4. compassionate
5. concerned for others
6. daring
7. dominant
8. easily upset
9. energetic
10. enjoys art
11. enjoys athletics
12. enjoys carpentry
13. enjoys fishing
14. enjoys hunting
15. enjoys poetry
16. enjoys power
17. enjoys vigorous activity
18. enjoys watching plays
19. feelings easily hurt
20. feels inferior
21. focused
22. forceful
23. gullible
24. home-oriented
25. independent
26. individualistic
27. interested in mechanical things
28. neat
29. orderly
30. poetic
31. precise
32. reckless
33. religious
34. risk-taking
35. self-reliant
36. self-sufficient
37. spiritual
38. tender
39. warm
40. wild

Scoring

You are now going to obtain scores for ten different factors, each score resting on your responses to four items. For the moment, we can refer to the factors as Factor A, Factor B, Factor C, and so on up to Factor J. You may want do the scoring on a fresh sheet of paper.

For Factor A, write down your numerical responses to items 4, 5, 38, and 39. Then total those responses. This will give you a score between 4 and 16 for Factor A. Once you done this, you can perform the same operation for each of the other nine factors. For each factor, be careful to record the numerical responses for the appropriate four items. Those items are as follows:

Factor B: items 8, 19, 20, 23
Factor C: items 10, 15, 18, 30
Factor D: items 3, 24, 33, 37
Factor E: items 2, 7, 16, 22
Factor F: items 12, 13, 14, 27
Factor G: items 6, 32, 34, 40
Factor H: items 25, 26, 35, 36
Factor I: items 21, 28, 29, 31
Factor J: items 1, 9, 11, 17

Interpretation

These items are drawn from a longer inventory, and I am sure it is obvious that any score resting on only four items is rather insubstantial. If we are interested in determining just how you compare with other people on each of ten different traits, we need reliable scores based on more than four items per scale. I just wanted to give you some food for thought before I tell you more about the research on which this inventory is based. But now, let us consider just what sort of personality traits these ten factors represent. As you consider them, you are obviously free to decide how well your scores reflect the way you actually see yourself.

I call Factor A **Nurturance**. The items are *concerned for others, compassionate, tender,* and *warm.* Related items in my longer inventories are *enjoys helping others, helpful to others, kind, responsive to others, sympathetic,* and *understanding of others*. It is easy to think of other possible names for this factor — e.g., compassion, social concern, relatedness, or social interest.

Factor B: **Emotional Accessibility**. The items are *easily upset, feelings easily hurt, feels inferior*, and *gullible*. Related items in my longer inventories are *cries easily, emotional, fearful, moody, needs approval*, and *needs security*. Other possible labels for factor B would include emotionality, emotional sensitivity, and emotional vulnerability.

Factor C: **Aesthetic-Imaginal Orientation**. The items are *enjoys art, enjoys poetry, enjoys watching plays*, and *poetic*. Related items in the longer inventories are *intuitive, enjoys reading novels, creative, contemplative, receptive*, and *perceptive*. An alternative label for the factor might be aestheticism, but something more in the way of openness to the inner realm seems to be involved as well.

Factor D: **Piety**. The items are *committed, home-oriented, religious*, and *spiritual*. Related items include *faithful, forgiving*, and *obedient*. Possible alternative labels for the factor would be spirituality, religiosity, commitment, and fidelity.

Factor E: **Ascendance**. The items are *bossy, dominant, enjoys power*, and *forceful*. Related items include *aggressive, a leader, assertive, controlling, hard-headed*, and *stubborn*. Other possible labels would be dominance, aggressiveness, and power orientation.

Factor F: **Concrete Action**. The items are *enjoys carpentry, enjoys fishing, enjoys hunting*, and *interested in mechanical things*. A related item is *interested in science*. This factor captures a pattern of interests found more often in men than in women, and it might also be called practicality or materialism.

Factor G: **Impulsivity**. The items are *daring, reckless, risk-taking*, and *wild*. Related items are *brave, courageous, fearless, fickle*, and *undependable*.

Factor H: **Autonomy**. The items are *independent, individualistic, self-reliant*, and *self-sufficient*. The factor could also be called independence or self-sufficiency.

Factor I: **Orderliness**. The items are *focused, neat, orderly*, and *precise*. Related items include *achievement-oriented, likes mathematics, logical*, and *rational*. The overall pattern suggests a quality that various writers have considered a masculine mode of consciousness, using such terms as Apollonian consciousness, patriarchal consciousness, and the logos principle.

Factor J: **Activity.** The items are *active, energetic, enjoys athletics,* and *enjoys vigorous activity.* Related items include *competitive, enjoys watching baseball, enjoys watching boxing,* and *enjoys watching football.*

Now consider the pattern of your own scores. On which of these factors do you score high? On which do you score low? Does this make sense in terms of how you generally think of yourself? Bear in mind that we are dealing here with self-description. Your relative standing on these factors might be somewhat different if we had scores obtained in some other way — say, scores based on observations of your behavior in everyday life.

We should note as well that the first four factors — Nurturance, Emotional Accessibility, Aesthetic-Imaginal Orientation, and Piety — are all factors of femininity. This is true in the sense that they are composed of items that have been judged as generally more descriptive of women than of men. Furthermore, women tend to score higher than men on these factors. The remaining factors — Ascendance, Concrete Action, Impulsivity, Autonomy, Orderliness, and Activity — are factors of masculinity. They are composed of items judged to be generally more descriptive of men, and men tend to score higher than women on at least five of these factors.

What about your own scores? Are the high scores predominantly on the masculine factors or the feminine factors, or are they as likely to go one way as the other? Since these factors are fairly independent of one another, it is actually unusual to see a set of scores for an individual in which all the masculine scores are higher than the feminine scores, or vice versa. There is certainly no particular virtue in achieving such a score pattern.

It is important to recognize as well that the difference between the average score for men and the average score for women is not great on any of these scales. The distributions overlap considerably. You are more likely than not to find that your higher scores point to a mixture of masculine and feminine traits, while your lower scores as well show a mixture.

THE UNDERLYING RESEARCH

Many questionnaires have been devised for assessing gender differences in self-description. In most cases, the test constructors have sought an overall measure of the extent to which a test-taker's responses are like those of men in general or more like those of women. The result in such a case has been a unidimensional scale in which one extreme represents masculinity and the other extreme represents femininity. In some questionnaires high scores represent a predominance of masculinity, while in others they represent femininity.

It is possible, of course, to be very masculine in some respects and, at the same time, very feminine in other respects. Recognizing that masculinity and femininity can vary independently and that they do not necessarily function in opposition to each other, some psychologists have taken a two-scale approach to measurement. What they have sought is an instrument containing both masculinity and femininity scales, each resting on a different set of items from the other. The work of Sandra Bem (1974) is a noteworthy example of this strategy.

In the course of my work with test development (Coan, 1989), however, I was more concerned with the fact that neither masculinity nor femininity is a unitary concept. Each represents a variety of traits in which men and women differ, and I wanted to capture and identify the major components. A few other people have done related work, applying factor analysis to the items in various instruments that were already in use.

I chose to start afresh with a reasonably comprehensive set of gender-differentiating items. In devising items, I took into account what was known about gender differences as well as the ingredients of various contemporary ideas regarding the nature of masculinity and femininity. I also considered longstanding cultural and mythic images of the masculine and feminine. The initial result was a set of 238 items, each consisting of a word or short-phrase descriptor.

These items were presented to the students in two psychology classes, who were asked to indicate for each item whether they considered it more characteristic of men or of women. Examining their responses, I eliminated items that did not yield substantial agreement. I also discarded a few items that the students found difficult to interpret. This left 200 items, which I administered to a sample of 539 students enrolled in introductory psychology. These students were asked to rate themselves on all items, indicating whether each item was not true, slightly true, mostly true, or definitely true.

I wanted to see what independent dimensions I could identify running through the responses of the subjects. Toward that end, I undertook three separate analyses, one based on the responses of just the male subjects, one based on responses of the female subjects, and one based on the responses of the entire sample. For each analysis, I calculated the intercorrelations of all the items and then performed a factor analysis. For each analysis, I extracted and rotated thirteen factors. Twelve of these recurred through all three analyses and seemed to represent clearly identifiable trait dimensions.

I reduced the 200-item inventory to an inventory of 110 items and administered it to another large sample of introductory psychology students. Again I performed three analyses and found the same twelve factors appearing again. Ten of these factors are the ones you have already seen — the ones we scored for the 40-item inventory. I should note, however, that two of these factors — Autonomy and Orderliness — collapsed into a single factor in the analyses for women and for the total sample. They were distinct in the analysis of male data, however, and I would argue that we still have substantial evidence of four dimensions of femininity and six dimensions of masculinity in the realm of self-description.

There is another qualifying note I must add. We started with items that were judged as either more descriptive of men in general (hence masculine items) or more descriptive of women in general (feminine items). For the most part, I also found that men were more likely to apply the masculine descriptors to themselves and women were more likely to apply the feminine descriptors to themselves. The gender difference, however, was not as clear-cut for self-ratings as for the group judgments obtained at the outset. A few items judged masculine were chosen as self-descriptors more often by women than by men.

In general, I did find that men tended to score slightly higher than women on the scales composed of masculine items, while women tended to score higher on the scales composed of feminine items. The one exception was the Autonomy scale. On the basis of the initial judgements — which you may regard as an indication of gender stereotypes — this is a masculine scale. Yet in the sample to which I administered the 110-item inventory, the women scored a little higher on this factor than the men did. The difference was not statistically significant, but the mere fact that it ran in the unexpected direction raises interesting questions. Was this just a chance anomaly involving this one sample? Assuming we would find the same trend in other samples, does this mean that the gender stereotype runs counter to fact? Or do university students represent an atypical segment of our society? Still another possibility: Is this evidence of an ongoing shift in the roles of men and women in our society — a trend involving increasing independence on the part of women?

I said above that twelve factors appeared repeatedly in my analyses. Ten of them were the ones we have already considered. What about the other two? They both involve items judged as either masculine or feminine, but neither factor as a whole can be considered a masculine or feminine factor.

One of these factors I called **Expressiveness vs. Reticence.** If we look only at the items most central to this factor, it appears to be a bipolar factor with a feminine quality at the high end and a masculine quality at the low end. This is consistent with the views expressed by writers like John Gray (1992). The picture is not quite so simple, however, for there are some items with moderate or weak loadings that run counter to this simple interpretation of the factor. Some masculine items that go with the Expressiveness end of the factor are *noisy, wild, conceited,* and *jokester.* Feminine items that go with the Reticence end of the factor include *passive, quiet, soft-spoken, timid,* and *sexually restrained.*

The other factor is **Sensuality.** It is neither a clearly masculine factor nor a clearly feminine factor. Some of the most representative items are masculine in terms of both social ratings and self-ratings: *lustful, promiscuous,* and *wild.* Others are feminine in terms of both social ratings and self-ratings: *romantic, affectionate, sentimental,* and *shares feelings.* Still others were judged more characteristic of women but received higher average self-ratings from men. These include *seductive, sensual,* and *physically attractive.*

IMPLICATIONS OF THIS RESEARCH

It should be clear from this research and from your own scores that it is not very useful to think of masculinity and femininity as the opposite ends of one grand dimension. There are different ways of being masculine and different ways of being feminine. Each of the trait variations we have noted has turned up before, either in research or in conceptual treatments of gender differences.

The dimensions of Nurturance and Ascendance have been recognized most often, in one form or another. The feminine factors of Emotional

Accessibility and Aesthetic-Imaginal Orientation and the masculine factors of Concrete Action, Impulsivity, and Autonomy have also appeared whenever appropriate items have been employed to represent them. It is important to note that all these factors, along with Piety, Activity, and Orderliness can vary independently of one another. Thus, a given individual can be high on both Ascendance and Nurturance, while low at the same time on Aesthetic-Imaginal Orientation and Impulsivity. You can characterize that individual as very masculine or as very feminine only if you focus on some traits and ignore others.

Our findings for Sensuality and Expressiveness vs. Reticence point to the additional possibility that a trait can have both masculine and feminine modes of expression. That possibility is not limited to these two dimensions. Thus, we may find ways of nurturing and caring for others that are more characteristic of men than of women. And women in leadership positions may display ways of directing action (manifesting Ascendance) that we do not often see in men who assume those positions.

Let us also bear in mind that while self-report inventories are a useful and convenient way of assessing a wide assortment of trait variables, they have definite limitations. People vary in ways that we cannot assess by this means. There are ways in which people differ in thinking, perceiving, imagining, and behaving that we cannot capture with self-description, either because people are not sufficiently aware of their own tendencies or because they have false impressions of those tendencies. There is an obvious need for additional forms of measurement.

4. Biological and Social Determinants

Some writers, when discussing differences between men and women, have argued that biological factors are all-important. Others have contended that learning within a given social environment is the key and that all major differences are products of the society in which the individual grows up. Yet an emphasis on either biological or social determinants to the neglect of the other is unwarranted.

All of our behavior either is a product of learning or is subject to modification by learning. Yet learning leaves its imprint in a brain that already has inborn properties that vary from one sex to the other and from one individual to another. Biological differences between men and women and among individuals affect the way any social input is received. Conversely, learning can have lasting effects on biological processes. Throughout life, there is an interplay between biological and social determinants.

BIOLOGICAL DIFFERENCES

With occasional exceptions, each of us begins life as a single cell with 23 pairs of chromosomes. One pair consists of the sex chromosomes, two X chromosomes in the case of the female and an X and a Y chromosome in the case of the male. The exceptions include the individual with only a solitary X chromosome (the XO case), the individual with two X chromosomes and one Y chromosome (XXY), and the individual with an X and two Y chromosomes (XYY).

In the early stages of development, male and female embryos look pretty much alike. They seem to be sexually undifferentiated. The original anatomic structures proceed to differentiate and assume the typical male and female forms under the influence of the so-called male and female hormones. The male hormones are called androgens, the most important being testosterone, while the female hormones are called estrogens. Each of us is both male and female in the sense that our bodies produce both sets. The difference is in the relative amounts of the two sets.

Testosterone begins to surge in the fetus at about eight weeks in utero, and normally the male body produces far more of this than the female body. Without this surge in the male, the body would develop into a form closer to that of the mature female. Indeed, it has been suggested that the default human form is female and it is only because of a surge in testosterone at a crucial stage in development that the typical male organs take shape.

At a later stage in fetal development, testosterone may have its most important impact on brain development. It is likely that estrogens also play an important role here, but the evidence so far is limited. Various male-female differences in brain development and function have been postulated, and there is some research evidence of such differences, particularly with respect to the hypothalamus, some parts of which appear to be larger in male brains while other parts appear larger in female brains.

Differences in brains may directly underlie differences in thinking and behavior between men and women. They may also mediate changes in other organs that in turn have behavioral effects. For example, secretions from the hypothalamus stimulate the pituitary gland to produce gonadotropins, which stimulate the ovaries and testes to produce estrogens and androgens. These in turn lead to behavioral effects by virtue of their impact on other parts of the brain.

There is no doubt that androgens and estrogens affect our emotions and behavior. Higher levels of testosterone are associated with anger and with sexual desire in both men and women. The various estrogens govern the changes that take place during the menstrual cycle and over the course of pregnancy. Increased secretion of the estrogen prolactin is necessary for milk production in the new mother. Prolactin is normally secreted by the pituitary gland in both sexes, and it is possible that estrogens play a role in nurturant behavior in both men and women.

Chromosomal and Hormonal Abnormalities

Various departures from the biological norm for males and females are possible. They tell us something about the role of hormones in our behavior and thinking, but they also raise questions that remain to be answered.

A case in point is the child with androgen insensitivity syndrome. If we have at conception an individual with an X and a Y chromosome, the sexual identity is clearly male. If the cells of the body do not respond to the androgens produced by the testes and adrenal glands, however, the male reproductive organs fail to develop in the fetus, and the external sexual organs of the newborn will appear to be female. The fact that the uterus and fallopian tubes are missing will not be so obvious. The child will be identified and reared as a girl. Even though the genetic sex is male, this individual will almost always accept the assigned gender identity without question. It is rare for the true situation to be recognized before puberty. At that stage of life, the child appears to be feminine and may begin to develop womanly breasts. The failure of menstruation to occur, however, may trigger a medical investigation that reveals the underlying abnormality.

An early sexual assignment counter to the genetic sex of the infant also occurs at times as a consequence of a surgical accident. Occasionally an inept physician intending to perform a circumcision will remove the entire penis. Recognizing the problems that the boy would face in the future, the doctor in consultation with the parents may opt for additional surgery to transform the external anatomy to a female form. The child will then be treated henceforth as a girl and in most cases will grow up accepting the assigned gender identity. There have been instances, however, where that assignment is rejected and the child later insists that he is, in fact, a boy.

A condition in sharp contrast to androgen insensitivity is the one known as congenital adrenal hyperplasia. Here we have an abnormally high level of production of androgens beginning in the prenatal period. In the case of the girl, this may not start at a sufficient level early in fetal development to prevent the normal development of the female reproductive organs. The external genitals, however, may appear somewhat virilized at birth. The clitoris may be enlarged and resemble a small penis, and corrective surgery may be performed. The child is recognized as a girl, however, and is unlikely to reject that identity. Studies of such girls, on the other hand, indicate that they tend to be more physically active than other girls their age and are more likely to be characterized as tomboys. Boys born with congenital adrenal hyperplasia also tend to be very physically active, but they run a risk of reaching puberty several years too early. These days both boys and girls with this condition are commonly treated with cortisone to correct for the overproduction of adrenal androgens.

A more puzzling condition is transsexualism. In this case, we see an individual who is genetically male and recognized as a boy at birth who decides in the course of growing up that he is really a girl, or an individual who is genetically female and so recognized at birth who decides that she is a boy. Male transsexuals are far more numerous than female transsexuals. As yet there is no satisfactory explanation for this condition. Some theorists have claimed that the condition stems from an unusual pattern of parent-child interaction in infancy, but there is no consistent evidence to indicate that the parents of transsexuals are much different from those in the rest of the population.

There is also no evidence of physical anomalies that would explain the condition. Transsexuals seem quite ordinary with respect to their levels of male and female hormones. It is conceivable that a hormonal irregularity at a late stage in fetal development would cause a male's brain to become feminized (because of a temporary drop in androgen production) or cause a female's brain to be masculinized (because of a surge in androgen). There is no research evidence to support this conjecture, however, and it flies in the face of the fact that there is no apparent association between congenital adrenal hyperplasia and transsexualism.

Sexual orientation presents the same kind of puzzle with respect to causation. Heterosexuality is as much a mystery as homosexuality. Early hormone exposure may be a factor, but it is not a decisive one. Girls exposed prenatally to excessive androgens are more likely than other girls to manifest a homosexual or bisexual orientation later in life, but most of them do not.

There is reason to believe that genetic factors are involved, but they have not been specifically identified. We know that the brothers of gay men are a little more likely than men in general to be homosexually oriented. Furthermore, the identical twins of gay men are still more likely to be gay. About half of them are. Studies of lesbian women show the same pattern. About fifty percent of their identical twins are also lesbians. Since identical twins are genetically identical, this would indicate that genetic factors are clearly involved but not all-determining.

Something else is involved. Yet despite various theories concerning family patterns responsible for homosexuality, there is no satisfactory evidence for an environmental cause. It is possible that, just as each of us at the beginning of life is potentially both male and female, each of us is potentially both heterosexual and homosexual, and subtle factors may tip the balance. It is conceivable that if we grew up in a suitable environment each of us would be homosexually or bisexually oriented. The source of our sexual orientation remains a tantalizing mystery, but its basic elements come into play early in childhood, and they are certainly not a product of rational decision. Since we label ourselves homo sapiens and we like to regard ourselves as thinking animals, we are puzzled when we experience powerful impulses evidently driven by an instinctual force we cannot quite identify.

Deviations from the normal pattern of two sex chromosomes should be able to tell us something about genetic determinants of gender differences, but it is not easy to make the desired inferences from these rare cases. People with the XO pattern (Turner's syndrome) and XXY pattern (Klinefelter's syndrome) have an assortment of physical problems, and their psychological traits reflect their efforts to deal with those problems and with reactions of other people to their physical traits. Men with the XYY pattern have most often been found and diagnosed in prison populations, but the mere fact that they are there may mean that they are not representative of the general population of men with that chromosomal pattern.

In the case of Turner's syndrome (XO), the child is clearly recognized as female. She may proceed normally through childhood, but is likely to be shorter than her peers. The odds are that she will fail to have menstrual periods in her teenage years and will be infertile, though there have been rare cases of women with Turner's syndrome who have become pregnant and successfully carried their offspring to full term. Women with this condition are prone to an assortment of cardiovascular disorders, and kidney abnormalities, thyroid disorders, and diabetes often bring them to medical attention.

The child with Klinefelter's syndrome (XXY) is recognized as a boy, but he is likely to have some problems stemming from his deviation from the physical norm. He may be of normal height or even taller than average, but his body is likely to appear somewhat feminized. His testes will appear much smaller than normal, and in adolescence his breasts may become enlarged. He is likely to be infertile, but that is not inevitable. He may be able to overcome some of his psychosocial problems with hormonal treatment and exercise to increase muscular tissue. Children with either Turner's syndrome or Klinefelter's syndrome are usually of normal intelligence, but they are more likely than most children to manifest learning difficulties. Robert Stoller (1968) suggests that men with Klinefelter's syndrome are somewhat more likely than other men to be gay or to be transsexual, though most of them are neither.

Men with the XYY pattern tend to be a little taller than average, but there seem to be no other unusual physical features. Undoubtedly most of these men go undetected because they lead very ordinary lives. They are discovered whenever genetic testing is done for any of a variety of reasons. It appears that these men turn up in disproportionate numbers in prison populations. Some people have inferred from this that the extra Y chromosome makes these men more prone to such "masculine" traits as aggression. Yet in only a few cases is their incarceration the result of acts of violence. Some research suggests that XYY men tend to average a little lower on intelligence tests than their peers, but they generally fall within the normal range. We would need additional research to determine how much they really differ in personality and cognition from the general population.

OBSERVATIONS OF OTHER SPECIES

Studies of other species can throw considerable light on the biological determinants of behavior. As we compare various species of mammals, it seem reasonable to say that the simpler the brain of the animal, the greater the extent to which its behavior is controlled by built-in neurological programming. As we move to animals with larger and more complex brains, the built-in programs do not disappear, but learning plays an increasing role in shaping the specific form of any behavior.

In rats the patterns of sexual behavior are well-defined, but it is clear that the neural bases for both the male pattern and the female pattern are present in every rat. Both patterns naturally appear at times in male rats, as well as in female rats. The choice of patterns is subject to hormonal control. In rats and other animals, a sufficient amount of testosterone at a time close to birth is evidently necessary if male sexual behavior is going to appear later on. If a male rat is castrated and treated with estrogens, he is likely to exhibit the female pattern up to and including assumption of the receptive coital posture. A female rat subjected to ovariectomy and treatment with testosterone may display no sexual behavior at all, but the testosterone is likely to lead to the robust activity more typical of the male. The effect of testosterone on activity level and on the rough-and-tumble play more typical of males has been found in other species.

In the absence of any experimental manipulation, homosexual behavior has been observed in dozens of mammalian species. What is most often reported is males attempting to mount other males, but females may be seen at times attempting to mount both males and females. In addition, some animals display a lifelong interest in others of their own sex. This has been found, for example, in a fair number of rams. It has also been found in female monkeys. So it is clear that homosexuality, both as an occasional behavior and as a persistent preference, is a natural variant in the behavior repertoire of animals throughout the mammalian class. From an evolutionary perspective, heterosexuality is more prevalent for the simple reason that it is essential for the perpetuation of the species.

It is possible to argue on the basis of naturalistic observation of other species that some of the major gender differences in humans have a biological basis, even if they are not rigidly determined by biology. There are obvious parallels in the animal world. Thus, in most, but not all, animal species, aggressive or violent behavior is seen more often in males than in females. And the parallels are most striking when we look at primates, in particular our closest relatives in the jungle, the chimpanzees.

Young male chimpanzees display competitive behavior. They seek dominance but may form coalitions as they struggle to establish a dominance hierarchy. The interaction of young female chimpanzees is more cooperative and more conducive to development of the support systems they will need when they begin to bear their own young.

We tend to think of nurturance in general as feminine or maternal. To be sure, lactating and nursing are within the province of the female members of the mammalian world. Yet nothing else in the realm of parental behavior appears to be manifested exclusively by female animals, even if the impulse to nurture the young is usually stronger in females. Males of various primate species can be seen to carry infants around, and they adopt infants that have been orphaned or abandoned.

Males of many species will go to great lengths to protect their young. Living in Tucson, I have often had an opportunity to observe many desert dwellers in their natural state. Some of them move about in family groups. Among birds, quail are an obvious case in point. I often see two adults strutting through the desert accompanied by a number of little chicks. As for mammals, I most often see a family of javelinas scurrying across a road, moving through a wash, or foraging among the assorted vegetation.

One day I sat in a restaurant on the outskirts of town. Through the window I saw a pair of adult javelinas in a desert area with a half dozen babies dining on food scraps. As they did so, a coyote appeared on the scene and began to creep toward one of the babies. The adult male of the family, however, took note and placed himself squarely between the piglet and the intruder. The coyote towered over his stout challenger for a moment but soon made a wise decision. He turned and departed from the scene.

The impulse to nurture can easily cross species lines. It is no accident that so many of us have pets. Those of us who are drawn to cats may be seduced by their little round faces and their proclivity for emitting a sound that resembles the cry of a human infant. Many of you have probably seen pictures of a gorilla who was given a kitten to play with and fondle, and you may have been struck by the extent to which the ape was visibly upset after the kitten died. Female cats have been known to nurse puppies, and female dogs have been known to nurse kittens. There have even been instances where human infants were adopted and nursed by wolves.

A dramatic rare case of adoption recently came to light in a California zoo, where the cubs of a mother tiger were all born prematurely and died. When the tiger lapsed into a depression, a replacement litter was sought. No tiger cubs were available, but a litter of orphaned pigs was found. Each was wrapped in a bit of tiger skin and placed with the tigress, and she quickly proceeded to nurse five porcine infants, and they in turn eagerly snuggled with their new mother.

THE INFLUENCE OF THE SOCIAL ENVIRONMENT

Whatever the constitutional makeup of the child, social factors begin to shape development in infancy. Recognizing that the child is either a boy or a girl, parents form various expectations regarding the future course of development. The child is given clothes deemed appropriate for either boys or girls and then gender-specific toys — perhaps dolls, perhaps miniature cars and trucks. The child may be encouraged to play certain games and to avoid others. The message that this is what boys do or this is what girls do is likely to recur in many different situations.

Much of the early behavior shaping is accomplished just by the environment that parents provide, with all the clothing, toys, play equipment, and books that the parents deem suitable. Along with this, the parents provide a stream of verbal messages from which children learn. At times the child is given a strong message when he or she does something the parents consider inappropriate, and punishment ensues. If the parent is very pleased with the child's behavior, praise and encouragement may follow.

Of course, no child starts out as a piece of clay to be molded any way the parents choose. Each infant begins with a unique set of potentials, and these are bound to differ somewhat according to the sex of the infant. Parents will react to whatever the child presents in the way of behavior. Parents do not treat boys and girls differently just because they are aware of the child's sexual identity. Boys on the whole tend to receive both more punishment and more praise than girls do for the simple reason that they tend to be more physically active.

The more education the parents have had, the more they are likely to believe that they will treat their sons and daughters alike and not impose strict gender-appropriate roles on them. Yet they will make distinctions unwittingly. The father will tend to identify with his son, while anticipating a certain affectionate relationship with his daughter. The mother will identify more with her daughter, expecting a somewhat different kind of bond with her son.

To some extent, children acquire ways of acting just by copying or imitating their parents. The parents are usually the child's primary role models in the early years. Identification facilitates this process. Just as the parent tends to identify with the child of the same sex, the child tends to identify with the parent of the same sex. The boy tends to see himself as like the father, and the girl tends to see herself as like her mother. If the boy feels at odds with his father, he may look for other male role models outside the family. The girl likewise may seek female role models other than her mother.

The modeling and imitation often have effects that no one in the family is aware of. I recall a nineteen-year-old roommate I once had in college. When I later met his father, I was struck by the extent to which the father's expressive movements matched those of the son — the facial expression, the stance, the tilt of his head, and the way he moved his arms and body when he walked. A few years later I was struck by the same kind of resemblance between a teenage boy who lived next door and his father. In both cases, I was initially inclined to attribute the resemblance to the genes passed from father to son. Yet in each of these two cases, I learned subsequently that I was not seeing the boy's biological father but rather the adoptive father with whom the boy had grown up.

Parent-child interactions in the early years are bound to have a lasting impact. Yet as the child grows, more influences come into play. The child meets and plays with other children outside the family, and their families differ in various ways from his own. He or she encounters teachers, friends of his or her own parents, and an assortment of other adults in the community. There are also possible role models to be encountered on television, in movies, and in books. In this way, the child becomes increasingly aware of the expectations of the surrounding society and of the alternative roles available within it. These expectations and roles vary somewhat from one ethnic group to another within our society, and they differ from those of other societies.

Cultural Variations

With respect to all gender differences we have considered, we find no absolute differences between men and women. Instead, we find much individual variation and overlapping distributions of traits. There are individual extremes that run counter to any given difference in averages. Some women are much more prone to violence than some men. Some men are much more gentle and nurturing than some women.

Given the variation within our own society, it is reasonable to expect still more variation when we compare people in different societies. In our society, the nuclear family has become the standard living arrangement in recent times. The household unit commonly consists of two adults, the mother and father, and one or more children. In an earlier period, a child might have grown up in close proximity with an extended family and would have had more frequent contact with grandparents, uncles, aunts, and cousins. Household tasks and parenting could then have been shared in ways not often possible today.

The impact of adults on children is subject to still greater variation when we consider societies where occupational role patterns and household arrangements do not conform to any of our own traditions. Such tasks as child rearing, cooking, housekeeping, manual labor, care of domestic animals, farming, and hunting can be assigned in various ways to men or to women. And imagine a society in which men sleep together in one place, while women sleep together in another, and all children grow up in close contact with mothers but not with fathers. The boy initially identifies more closely with the mother but is expected to take a big step toward manhood when he is roughly initiated into male society at puberty.

Bronislaw Malinowski (1927) has described a matrilineal family pattern among the Trobriand Islanders that differs considerably from the patrilineal pattern of European and American society. In that society, following sex play in childhood and promiscuity in adolescence, young men and women usually settle into monogamous living arrangements as adults. The couple may have children, but the father's biological role as progenitor tends to be unrecognized. According to Malinowski's description, men tend to assume dominant roles, but the father's role in relation to his children is that of a benevolent playmate. The boy grows up asserting authority over his sister and over her children when he reaches adulthood. As a child he recognizes his uncle as the man of authority in the family.

The work of Margaret Mead (1935) is most often cited as evidence that the gender-role differences to which we are accustomed are dictated primarily by culture rather by biology. She described three Samoan societies that contrast sharply with our own. One is the Arapesh society, where aggression and violence are discouraged, and both men and women are expected to be warm and nurturing, treating others in a gentle and caring fashion. In the Mundugumor society, according to Mead, we see just the opposite. Both men and women are typically very aggressive and often violent. In a third tribal society, the Tchambuli, Mead saw a typical pattern of male-female interaction that is the reverse of the one that we regard as more typical of our society. The woman in a marital relationship is the dominant and impersonal managing partner, while the man is more emotionally dependent and assumes little responsibility.

Mead sought to build a case for cultural relativism. She recognized that basic temperament varies from one individual to another and that the same range might be found in any society. Yet it would be subject to cultural conditioning. A tendency toward violence would be suppressed in Arapesh society and accepted or even encouraged in Mundugumor society. More important to Mead, however, is the point that such traits as aggressiveness and passivity are not sex-linked. She contended that the difference we see between men and women is a product of cultural conditioning, not of biology.

Mead provided many ideas worth pondering, but I think most social scientists would agree that she overstated the evidence for the general point she was trying to make. Others who have studied Samoan societies do not find the evidence as clear-cut and compelling as Mead would have us believe. It is still reasonable to believe that some of the temperamental difference between men and women is biologically based and that it arose through evolution at a time in our ancestral past when the care of infants necessarily fell primarily to women and the procurement of food by hunting fell primarily to men. On the other hand, there can be no doubt that gender differences are shaped by cultural conditioning and will vary from one society to another. In most societies, cultural pressures are likely to be consistent with the natural biological differences. They may tend to augment those differences or to minimize them. They are less likely to produce a wholesale reversal of the gender differences we most often see. In any case, both biology and culture have roles to play, and we have nothing to gain by exaggerating the importance of either to the exclusion of the other.

To this point, I might add another: neither culture nor biology remains fixed. Evolution comes to a stop only when a species becomes extinct. Survival of our own species no longer requires all men to be aggressive hunters nor all women to be nurturing tenders of infants. Perhaps the biological divide is gradually diminishing, and at the same time our society is certainly changing. The freedom of American women to assume virtually all the roles claimed only by men in the past is on the rise. Biology and culture may well conspire to reduce all the gender differences to which we are accustomed.

5. Guiding Myths

In the preceding chapters we have dealt with a sense of identity primarily in terms of gender identity, the sense that one is a boy, girl, man, or woman. Yet gender identity is only one facet of an individual's total identity or self-image. That total identity encompasses all the characteristics that make us distinct persons. It includes all the features that we feel we share with various people and groups, as well as those that make us different from anybody else.

At the core of our sense of identity is a central image, our picture of who we are. This is a distinctly personal or individual image, but it contains collective elements, elements derived from the mythic images that underlie the ways in which people in our society view themselves and others. We project these mythic images onto public and historical figures, as well as applying them to others that we have actually met and known. The prevalent mythic images in our society differ to some extent from those of other societies, but they have ancient roots. In some cases, we can trace them back over many centuries and observe the changing forms of their expression. Some of these images have to do with what the society regards as ideal masculine and feminine qualities. Growing up in a particular place at a particular time, it would be difficult to avoid thinking of yourself in terms of these images and developing a personal self-image that reflects the extent to which the collective images apply to you.

MYTHIC IMAGES OF MASCULINITY

Some of our basic ideas and images of the ideal man and ideal women can be traced to the Middle Ages, and others clearly have roots that we can discern in more ancient Greek, Roman, or Middle Eastern writings. An example is the image of the perfect knight, which was well articulated in medieval writings. The knight adhering to the codes of chivalry manifested such virtues as honesty, perseverance, loyalty, strength, honor, and self-denial for the sake of serving a worthy cause.

The knightly ideal was a standard of excellence applied to members of a warrior class in medieval society, and there was no expectation that it would be met by the far more numerous peasants, merchants, and craftsmen of that time. Yet while medieval knights have disappeared, the image has lived on and found other forms of expression. We see it in the codes governing duels in more recent centuries, and the knight has reappeared as a hero in Western novels and movies. The cowboy myth has little connection with the historical reality of American expansion to the West in the latter half of the nineteenth century. Yet the great hero has captured our imagination. He rides into town, rescues the honest folk from the bad guys, then rides off again into the sunset, his only reward being the knowledge that he has served a good cause. Now we are prepared to leap into space and into the future, where the knight will appear once again as the Jedi warrior of Star Wars fame.

George L. Mosse (1996) has noted as well an image of masculine beauty that was emphasized in Europe in the eighteenth century. Artists at that time resurrected the image they found in Greek sculpture of an earlier age. This physical ideal has found expression in more recent times in our views of body-builders and athletes. At times a preoccupation with masculine beauty has appeared in the writings of racist theorists, who see the physical ideal as representative of their own ethnic group and ugliness as characteristic of groups they despise. The Nazis of Germany, playing fast and loose with history to proclaim their special "Aryan" heritage, are only one case in point.

In contemporary American society we hear more about female beauty than about male beauty. A greater focus on male beauty would not have been surprising in an age in which women were expected to remain in the shadows of men and not attract attention to themselves. It is debatable whether the introduction of beauty contests for women represents a very significant advance for women. It obviously does not if it means that an attractive surface is the only quality in which we expect women to excel.

For men, physical beauty has usually been combined with other qualities in an image of manliness. The other qualities include physical strength and skill, courage, and assertiveness. Mosse notes the various ways in which this image has been idealized in artistic, athletic, military, and political contexts in Germany, France, and England, from the eighteenth to the twentieth century. He also notes recurring counter-cultural movements that proclaim the right of individuals to express themselves in ways that run counter to the "manly" ideal. Toward the end of the nineteenth century, there was a movement commonly known in Europe as the decadence, espoused notably by gay intellectuals who openly flaunted their homosexuality and adopted modes of movement and attire that were generally viewed as feminine.

Europe and America were not ready at that time to abandon traditional views of appropriate masculine and feminine behavior. In the last three or four decades, however, we have witnessed a wider cultural shift. The women's movement, in its contemporary form, has spread world-wide and begun to bring about major advances in the rights of women. Homosexuality has gained increasing acceptance as an alternative lifestyle. Male entertainers like Michael Jackson, David Bowie, and Boy George have become popular while projecting an image somewhere in between those that we have considered masculine and feminine in the past. And James Dean has remained one of the most enduring idols of the movie screen. He was only 24 when he died in 1955 after making just three movies, but we remember him as a sensitive, emotional young misfit, seeking a love and acceptance that lay beyond his grasp — hardly an embodiment of the tough, aggressive image of manliness.

We have now considered ideal images of masculinity in forms derived from medieval codes of chivalry and in ideals of "manliness" that can be traced to eighteenth-century art and more remotely to classical Greece. Another image worth noting is that bound up in what Max Weber (1958) called the Protestant ethic. He considered the Protestant ethic a key factor in the rise of capitalism following the industrial revolution, but he believed its basic ideas could be traced to the Reformation of the sixteenth century, particularly the version advanced by John Calvin.

Calvin embraced a doctrine of predestination. According to this doctrine, some people were elected by God to go to heaven while others were not. If you were among the elect group, you were destined to succeed in all your endeavors. If you were not so chosen by God, no good deeds or acts of faith would make much difference. If you were among the elect, you would be likely to acquire wealth, and you would display qualities that would be conducive to acquiring wealth.

I believe the wealthy already exert too much influence on affairs of state in our society, and I shudder to think of heaven as a place reserved for them, while people who have lived in poverty are denied access. Weber, however, was more interested in lauding the capitalist system than in painting a picture of heavenly rewards. He believed that people succeeded in this system when they were not motivated by excessive greed but inclined to live like good Puritans. Their lives were guided by such qualities as frugality, hard work, planning, self-denial, and strict control over passions — all qualities that Weber believed were conducive to steady success in the business world.

I am not convinced that the capitalist system works quite the way Weber believed. Of course, the Protestant ethic may have played an initial role in the rise of capitalism, and it still embodies one ideal operating in our society. I see this as an ideal masculine image because Weber was writing at a time when the capitalist world was clearly a world of men and women were expected to remain in a less conspicuous, domestic niche.

MYTHIC IMAGES OF FEMININITY

Some of the dominant themes that have recurred in our society's views of the feminine ideal can be traced back over a period of at least two thousand years. Some are rooted in early Christianity. Christianity began as a minor Judaic sect, but as the movement spread to the West it blended ideas derived from three major patriarchal cultures — Jewish, Greek, and Roman. It was certainly not a unified religion at the outset. Dozens of gospels were written in the first two centuries, and most of them were suppressed by the early church as it gained increasing power. Christians were recurrently subject to persecution by the Roman government, but in the fourth century under the emperor Constantine, Christianity became a state religion. The leaders of the church proceeded to select four gospels for inclusion in the New Testament, embracing a few theological and historical ideas that the early followers of Jesus in Jerusalem would probably have considered strange.

Two feminine images played prominent roles in these early texts, those of Jesus's mother Mary and Mary Magdalene. In the former, we are presented with the picture of a woman who is so free of sin that she remained a virgin throughout her life, though married to Joseph. References in the gospels to siblings of Jesus were construed by church leaders as referring to children of Joseph from an earlier marriage. Thus, Mary bore a son of God while leading a life totally devoid of sexual interest or expression.

The synoptic gospels suggest that Mary Magdalene was close to Jesus, but an early pope equated her with a Mary who appears elsewhere in the gospels and declared that she was a fallen woman who was redeemed by her devotion to Christ. There is no justification for the equation, since the name Mary (i.e., Hebrew Miriam or an Aramaic derivative) was a common name in the first century. The pope was simply applying a pre-existing mythic theme to underscore the importance for women of renouncing sin and confining themselves to lives of spiritual devotion and piety. More than one historian has argued, however, that Mary Magdalene was in all likelihood the wife of Jesus.

Mary Magdalene assumes a still more prominent role in some of the gnostic gospels suppressed by the early church. She is depicted as the chief disciple of Jesus and the one who understood his message better than anyone else. Some of those gospels also speak of Sophia, a goddess of wisdom. Unfortunately those texts did not meet with the approval of church leaders intent on keeping women in a subordinate position.

As Christianity spread northward in Europe, it came into conflict with pagan traditions among the Germanic and Celtic people that accorded greater freedom to women. To many Christians, it was hard to accept the possibility that women might engage independently in any sort of magical, spiritual, or healing ritual. Women who appeared to be acting that way were suspected of being witches, and witches by Christian definition were evil. Witches began to appear in folk tales as wicked women. Over a period of several centuries starting in the late Middle

Ages, numerous women suspected of witchcraft were burned at the stake in Europe or hanged in New England.

In early Christianity, the quiet virgin and the harlot represented the extremes of good and evil in women, and this basic polarity has continued to appear over the centuries in varying forms. In the Romantic literature of the nineteenth century, the good woman was pure, chaste, tender, and lacking in worldly wisdom, hence in need of control and guidance from men. The woman who did not conform to this image was the temptress, the femme fatale, who would seduce men in order to destroy them. Despite pressure to conform, however, some women have recurrently demanded their right to act independently — even their right to vote. Such women appeared in the decadence, and they appeared between the world wars. Cigarette ads portrayed only men smoking, but a young woman might suggest that the man blow a little smoke her way, so that she might enjoy a faint scent of the burning weed. Yet there were women who believed they were fully entitled to blacken their own lungs — exemplified by the flapper, who cut her hair short, dressed in a boyish manner, and could be seen with a cigarette dangling from her lips.

Perhaps the flappers were just individual women proclaiming their right to be different, but feminism as a movement proclaiming the right of all women to share some of the power and the roles that had been reserved for men has been around since the middle of the nineteenth century. It was not until the second half of the twentieth century, however, that the movement drew an unprecedented number of adherents and began to achieve most of its major successes.

Betty Friedan (1963), a major voice of the movement, drew a lot of attention with her book *The Feminine Mystique*. Friedan noted that Americans had come to idealize a life for women centering around homemaking and childrearing. Young women would see their major goal in life as acquiring a husband, having four children, and enjoying a leisurely existence in a nice house in the suburbs. All sorts of books and magazines were produced to provide women with the guidance they needed to snare a husband, to deal with problems in marriage, and to clothe, feed, and discipline children. Furthermore, the post-war era brought technological developments that drastically reduced the drudgery in housework. Yet the big problem, as Friedan saw it, was that too many women ended up feeling that their identities were defined by their husbands, children, and beautiful surroundings. As a result, they felt a desperate need for something more. They lacked a clear sense of who they were as individual people, and their days were filled with tasks that seemed meaningless.

Some critics have contended that Friedan's analysis applies only to women at the upper economic and educational levels in our society. It is true, of course, that women in the lower economic strata are more likely to work outside the home out of necessity, and their work may consist of unsatisfying chores. It is also true that many men in the lower economic strata are employed in menial jobs that provide little or no personal satisfaction. In any case, even if Friedan's analysis applies to only one segment of the women in our society, the problem experienced by these women is an outgrowth of a broader societal tradition that accords less freedom to women than to men in the choice of roles.

Madonna Kolbenschlag (1979) addresses the contemporary problems of women by discussing traditional folk tales that reflect the fantasies and aspirations of girls growing up in our society. In such tales as Cinderella, Sleeping Beauty, and Snow White, the heroine must pass through a phase of total passivity. It is either a period of confinement to menial toil or a prolonged period of sleep. The girl can do nothing herself to end this. She must wait to be rescued by the assertive prince of her dreams.

Kolbenschlag believes that industrialization brought an increased role division between the home and the workplace, hence a greater split between the working roles of men and women. The role of the married woman tended to be become one of idle dependence. She sees a recent trend, however, toward a "symmetrical family," where the role division between men and women is not so great. Ideally in such a family both partners are free to pursue meaningful work outside the home and then to share the domestic duties of meal preparation and childcare. She also notes that many people, male and female, may find greater satisfaction living single lives and remaining unmarried.

PERSONAL MYTHS

Each of us begins to develop a sense of identity, or a personal myth, early in life. This contains features of the people with whom we identify and elements derived from the reactions we get from other people and from the labels they apply to us. Like the people around us, we are also affected by the masculine and feminine mythic images prevalent in the surrounding culture, and we are bound to incorporate some elements of these into our own personal myth.

Our personal myth develops and expands as we grow through childhood. It evolves as we acquire more information about people and about ourselves, as we meet more people, and as we enter school and pass through the grades. It changes as we become more aware of our feelings, intuitions, abilities, and dreams.

In our adult years, we continue to experience changes in our bodies, in our work, in our families, in other people, and in the society and world in which we live. It is natural for our personal myths to evolve in response to these changes. We always tend to act in accordance with our personal myths, and these myths can provide a sense of meaning to our lives. Unfortunately, however, we often cling to myths that have ceased to be useful. They may have facilitated adjustment to our life circumstances at a given time or in a given stage of our lives, but they now leave us stuck in patterns of thinking and acting that interfere with our ability to cope with changes in our bodies, our work, our family situations, and our social milieu. When we find that we are stuck in a pattern that has become a source of pain for ourselves and others, we need to examine the underlying myth that is guiding us. The key to our freedom may lie in finding new myths and the pathways in life that they can open up for us.

PROBLEMATIC PERSONAL MYTHS

The quality that Betty Friedan called the feminine mystique was a central feature of the personal myths of the women in her Smith College graduating class. These were women who had sought fulfillment through childbearing and homemaking. Many other pathways have also become available for women, and any one of these affords the possibility of both problems and satisfactions.

Norma Jeane Mortenson was born in 1926. She was with her mother for only brief periods and spent most of her childhood passing through a succession of foster homes, unable to depend on any adult for a long period of time. She entered the first of three marriages at 16. She began posing as a model by age 20. At that age she was divorced, but she secured a movie contract and changed her name to Marilyn Monroe.

She won acclaim as an actress, and she had a unique talent for projecting both on and off the screen an image that combined sexuality with childlike innocence. On the movie set, she was usually late in arriving, and her erratic behavior often upset directors and fellow actors. Despite the rewards of money and fame, she seemed insecure and professed a lack of understanding of the sexual image that so many people ascribed to her. It is unclear to what extent she identified with that image, but her self-esteem evidently depended on receiving constant validation from others, rather than on a secure sense of her own worth. At 36 she died from an overdose of sleeping pills, and we shall never know whether she had deliberately chosen to commit suicide.

We often tend to envy people who have achieved great fame, but we frequently encounter evidence that many of them lead troubled lives. It seems that the possibility of achieving fame and then the cost of living up to it once one has achieved it can leave the individual clinging to a guiding image, centering around one salient trait, that cannot evolve as one's needs change over the years.

Robert Schumann was well established as a composer and music critic when he met Johannes Brahms. Seeing and hearing the compositions of the younger man, he declared that Brahms was surely the man destined to wear the mantel of Beethoven — a prospect that Brahms found intimidating. Brahms did achieve fame, but he produced many compositions that we shall never hear because he destroyed them. He did not want anyone to hear them because he was afraid they were not good enough for a man worthy to be Beethoven's successor.

Jascha Heifetz was known during his adult years as the world's greatest violinist. It is said that he mastered the Mendelssohn E-minor Concerto at the age of 7. Someone said at the time of his death that there had never been another violinist on a par with Heifetz and never would be. Perhaps it is unfortunate that Heifetz had acquired a unique reputation for perfection, because he became increasingly reluctant to perform in his later years. Someone said that if Heifetz played for an audience of one thousand people, he felt he was standing in front of one thousand people waiting for the great Heifetz to make a mistake.

If there was a pianist whose reputation was comparable to that of Heifetz, it would have been Sergey Rachmaninoff. As a young man in Russia, he aspired to be both a great composer and a great pianist. When his first symphony was ready for public performance, he received strong encouragement from Tchaikovsky, but Glazunov, who was to conduct the orchestral work, was much less impressed and put little effort into the performance. When the public performance met with a cool reception by the audience, followed by the caustic comments of critics, Rachmaninoff fell into a lengthy depression. After a considerable period of hypnotherapy, he was able to resume work, but a life of serenity lay beyond his grasp. He has since earned fame as a late Romantic composer, and during his life he was often dubbed the world's greatest pianist. At fourteen, I was privileged to hear him in a recital during his last year of life. I shall always remember him as a tall man with long slender fingers, a keyboard skill I could not hope to match, and a face that bore a look of perpetual sadness.

Few of us achieve great fame, but we may still cling to a guiding image that serves our needs quite well at a certain stage in life. It brings wanted attention from other people and enhances our feeling of self-worth. It may be the image of the strong man, the Don Juan or great lover, the star pupil, the life of the party, the great actor or stand-up comedian, the spreader of joy and sunshine, everyone's caretaker, or the outstanding mother of infants. As the body changes, as people around us mature, and as our social milieu becomes transformed, the image may become dysfunctional. Then we must let it evolve or replace it with an image that suits our new stage in life.

Sometimes we invest great energy and hope in a secret identity, an image at the core of our fantasies but one not evident in our behavior and quite discrepant with the way we are perceived by others. In reality, every young man is just an ordinary Clark Kent or Peter Parker, but some of us sense that inside we have great powers. We are Superman or Spider Man, and we must not let other people know of this identity. Or perhaps our secret identity, as a man or woman, is that of a great genius, or an artist who will one day be famous when people finally understand our work, or a woman of incredible beauty that the world has not yet appreciated. This kind of image may help us survive a period of great stress or rejection, but we need to outgrow it because the clash between the inner image and the outer reality is an additional source of stress.

I hesitate to tell you this, but in my mid-teens I managed to cope with the agonies of adolescence because inside I was a misunderstood genius whose intellectual and musical gifts would surely be recognized some day. How did you deal with that period in your life? Were you a carefree member of the popular elite crowd, an athlete, a proud scholar, or another one of us social misfits? What was your guiding image back then, and how has it changed since?

A guiding image that depends very heavily on a relationship with one specific person can also present problems. In my first year of college, I enrolled in a physics course. I remember a brilliant fellow student in the class who appeared headed for a career in physical science. One day he vanished from the class, and I learned his fate when I read the newspaper several days later. He had been in love with a young woman who became engaged to someone else and was soon to wed her fiancé. Deciding that his intended career would mean nothing if he could not be with his beloved, he committed suicide. Perhaps most of us can understand his pain. We have fallen in love and felt devastated when the partner we desired rejected us or chose someone else in our stead. In such a case, we may have to endure a fair amount of pain, but it is usually possible to survive and find fresh ways of loving, living, and undergoing personal growth.

When I think of Weber's Protestant ethic, I think of my maternal grandfather, a man who certainly lived by Calvinist principles, though he never became wealthy. Born in Arkansas in the latter part of the nineteenth century, he had no more than an elementary school education, but he taught himself many things and could express himself in simple prose that had an almost Lincolnesque quality. Intent on working alone and being his own boss, he decided as a young man to be a carpenter, while doing a bit of farming on the side. He was always a man of his word and duty-bound to church and family. Yet he always maintained a cold surface, giving to others only what he knew was due according to the ethical and religious codes by which he lived. When he built houses, he was harsh and demanding with the men who worked under him.

As a child, I could never feel close to my grandfather, but I vividly remember a scene in his later years. My mother, brother, and I went to visit my grandmother, who lay near death in a hospital bed. My grandfather sat by her side as he spoke to us. He reported that my grandmother had told him that she thought he did not love her anymore, and this had touched him very deeply. He had come to realize that he had hurt her with his cold, insensitive manner and that he had also hurt men who had worked with him. At that moment, I witnessed a man in the process of transformation. In the ten years remaining until his death at 75, I saw a man who remarried and who manifested a warmth and caring for other people that had never been apparent before. I suspect he was aware of a shift in his own sense of identity. As I think of him, I am reminded of the reformation of Ebenezer Scrooge in Dickens' *A Christmas Carol*.

6. Major Mythologies

In the preceding chapter, we considered prominent mythic images that underlie common ideas in our society regarding the nature of men and women. We now need to take a broader look at mythic images by considering the gods, goddesses, and heroes who have assumed prominent roles over time in the major mythologies of various societies.

To some extent, these divine and heroic figures mirror the societies in which they appear. Thus, among them we find husbands, wives, kings, servants, artists, and craftsmen. Yet the more prominent figures may persist over a long time in the thinking of ordinary people while major changes are taking place in the society itself. Furthermore, we can find some basic patterns running through these divine and heroic figures that transcend the boundaries of any one society. Turning from one major mythology to another that has arisen in a different part of the world, we find certain images and themes recurring.

In view of these recurring themes, we have reason to believe that mythologies can be a fruitful source of insights into universal patterns that underlie the ways in which we experience the world, ourselves, and other people. To some extent, these patterns are inborn. They are an expression of the archetypal roots of human thought, perception, and feeling.

We can gain valuable insights into the flow of our lives if we examine the ways in which we respond to various archetypal images. We can think of each image as a potential way of acting, feeling, or thinking. If you are open to this kind of exploration, you would ask as you encounter each image whether there is something familiar about it. If so, in what way have you encountered it? There are several basic possibilities:

(1)   You may recognize it as something in yourself. It may be something you clearly identify with and consider a part of your basic nature. Or you may simply recognize it as a tendency to which you may or may not give full expression. Perhaps it is a quality you want to acquire and cultivate or one that you would like to outgrow and leave behind. For example, the image with which you identify may be that of a leader, a caring friend or mother, a wild man or woman, a teacher, an artist, an athlete, a beauty, a combative warrior, a great lover, a clever trickster, a saint, or a person of great emotional sensitivity. Does any of these images represent something you see in yourself or something you want to develop?

(2)   On the other hand, you may recognize the image as representing something you would like to find in someone else, someone to whom you could relate in a significant way. It might be a quality you would hope to find in a lover or mate. Perhaps it is a quality you would value in a close friend, in a mentor who could provide you with guidance, or in a younger person to whom you could offer guidance.

(3)   You may recognize the image as representing a quality that you have perceived in other people and found disturbing. You may have seen it in people you have known or met, people to whom you have reacted with anger, fear, or disgust. Perhaps you have perceived it in public figures, fictional characters, or people you have heard of in the news, people you would rather not encounter.

(4)   It is also possible that the image represents a quality that you have recognized in other people, but one that evokes no strong feelings. You do not see it in yourself, but it does not matter whether it is manifested in someone else.

As long as we are alive, at any stage in life, we can seek further personal growth. As I suggested in the preceding chapter, one key for doing this is to examine the guiding myths that are operating in your life and seeing whether they are really serving your needs. In the recurring images found in world mythologies, you can find potential modes of living that you may find valuable to explore and cultivate.

SOME GENERAL OBSERVATIONS

It may be impossible to gain a truly comprehensive picture of any major mythology. Mythologies are constantly evolving, and the gods, goddesses, and heroes that appear prominent at one time may be quite different from those of primary importance earlier or later. Furthermore, the information we obtain is often inaccurate. It tends to be distorted because of the biases and incomplete understanding of people from outside a given place or time who serve as our source of information. Christian writers have tended to look for concepts corresponding to their own god when writing about Taoism and Buddhism. In an earlier period, Romans seeking to understand Celtic myths tended to translate Celtic gods and goddesses into Roman ones. To learn about mythologies of the past, of course, we are dependent on available written records. The best records of Celtic myths are those recorded by Gaels in Ireland. Our knowledge of the related myths of Celts in other parts of Europe is more fragmentary.

One reason for constant change in myths is that neighboring people interact in various ways. A deity important to one group may be borrowed by another or assimilated to an existing deity, who then takes on properties of both figures. Or the borrowed deity may take on roles in the new setting to accommodate the figures already revered. It is also possible for a god or goddess prominent in one community to become pre-eminent over a much wider territory, particularly if the people of the original community manage to extend their rule over a wider area. On the other hand, if ruling authorities try to demand allegiance to a specific form of their favorite god or goddess or to the name that represents that mythic figure, they are likely to have only limited success. People may utter the required oaths of allegiance, but their words soon ring hollow, for they will cease to carry any emotional resonance with the lives of those who utter them.

Even without the interaction of different communities, mythic figures will constantly evolve. They will fluctuate in importance. A god preeminent at one time may be replaced by his offspring and cease to receive any attention. At any one time, a mythic figure may assume a variety of forms, appearing in different guises depending on the situation. We may find a shift from the mortal to the immortal, the heroic human figure at one time becoming a god or goddess. A shift between male and female is also possible. Kuan Yin, an important figure in Mahayana Buddhism, is a case in point. In recent centuries, Kuan Yin has usually been portrayed as a goddess, but at an earlier time depictions of Kuan Yin as a male god were common. Kuan Yin remained an embodiment of divine compassion in both forms.

Sometimes there are marked changes in the character of the figure. A god or goddess may change from being a protective figure to being a destructive one, and it is possible for these two sides of that figure to alternate in rapid succession. In India, Kali is viewed primarily as a frightful goddess of destruction, but she also represents the joy and peace that come with release from the realm of the transient forms that have been destroyed. Over time, some changes result from changes in the mores, outlook, and needs of the community. A deity may shift in character and assume a novel role. In classical Greek mythology, Artemis was a lone, virginal, self-sufficient huntress. In the city of Ephesus in the first century, however, she was worshiped as a many-breasted mother goddess, who could ensure an abundance of crops in the region surrounding that community. It is possible that in that part of western Asia Minor, a mother goddess of different origin just happened to be given the same name as the virginal Greek huntress.

THE RANGE IN CONTENT OF MYTHIC IMAGES

People create mythic images in the course of their efforts to comprehend everything about them that touches their lives. The various forces and features of nature are given expression in myth. Thus, there are mythic images to represent the sun, the moon, the sky, air, earth, water, rain, animals, plants, fertility, birth, and death. The more people are aware of their dependence on a particular feature of nature, the more central its place in myth. Solar deities abound in mythology, for we can hardly ignore the heat and light of the sun nor their absence at night, and our lives are affected by the changes from one season to the next. Special attention may be given as well to the nearby river, with its fluctuating flow, and to any animal or plant that serves as a primary source of food.

The human qualities that we experience in ourselves or perceive in others will also be represented in the deities and heroes of mythology. We can expect to find mythic figures who are compassionate, protective, destructive, frightening, tricky, commanding, or childlike, and any of these qualities may be attributed to a figure representing a feature of the natural world.

The customs, occupations, social structure, and gender roles of the worshiping society are also likely to be mirrored in the mythic realm. A society that relies on agriculture will differ from one that relies on hunting. The form of government of a society and the kind of interaction it has with neighboring societies will affect the mythology. The prominence of kings, queens, and warriors will vary from one mythology to another.

Many roles of gods and goddesses simply mirror those within the community. Thus, gods may govern and engage in warfare, while goddesses give birth, care for the young, and exercise domestic skills. Yet it is difficult to find any one role that is exclusively exercised in all mythology by deities of just one sex. Agriculture may be the domain of either gods or goddesses, and so may ruling, hunting, and warfare. In some mythologies, a god and his female consort jointly serve the same function, or at least complement each other in serving that function. It is also possible for a god or goddess to possess features of both sexes or to appear as male at one time and as female at another time.

A mythology will also reach beyond what we have recognized in the world about us. Mythic figures can be expressions of qualities we hope to realize in ourselves or hope to find in others, human potentials we sense as possible but not yet seen. The figure may be uniquely compassionate, may have a gift for intuition, a special understanding, or a skill in resolving social conflicts. In a time of oppression, an awaited savior may assume a prominent role.

A mythology can also include images that provide a sense of connection to the mysteries of our existence that lie beyond the reach of our understanding. The complex brains that we have developed in the course of biological evolution have their limits, and it is well to realize that we shall always be surrounded by mystery. The mystery pertains both to the nature and origin of the universe itself and to the ground of our existence

as sentient beings. Psychic phenomena suggest that we are all intercon-
nected on some level more basic than the level on which we experience
ourselves as isolated individual islands of consciousness, and perhaps there
is a universal ground of consciousness that extends beyond the narrow
scope of one species on the planet we call Earth.

THE EARLY ROOTS OF MYTHOLOGIES

When we compare the earliest known forms of mythologies of various
groups who speak Indo-European languages, we find marked similarities.
Among the early speakers of Germanic, Celtic, Greek, Romance, Slavic,
or Indo-Iranian languages, we find tales of major struggles between two
classes of superhuman beings. On the one side are beings representing
the forces of light, day, fertility, wisdom, virtue, and life. Pitted against
them are beings representing night, darkness, evil, and death.

We also find some evidence of names that have common linguistic
roots when we compare these mythologies. It seems reasonable to specu-
late that these mythologies have evolved alongside the languages in which
are expressed. If we could trace all these mythologies back eight thousand
years or so, we should find them converging on one common source, the
mythology of those who spoke the early Indo-European tongue from
which such languages as English, Russian, Spanish, Hindi, Farsi, Gaelic
and numerous others all stem. We have no written record of that early
language. The best we can do is attempt to reconstruct it following recognized linguistic principles. Perhaps we could attempt a comparable
reconstruction of the early Indo-European mythology, but that might
require a bit more imagination.

As we look at early Greek, Celtic, Germanic, Hindu, and Persian
myths, we see a recurring pattern of conflict between forces represented
by beings who are predominantly male struggling for power and control.
It is reasonable to seek a common source, an earlier form of the great
cosmic struggle. Yet there is something more that we can expect to find
as we scan back over past millennia. Over a wide territory, extending at
least from the Pyrenees to Siberia, we have found stone-age sculptures
and cave paintings that focus on images of the female body. We find both
figurines and drawings of female figures with enlarged breasts and bellies
that clearly suggest reproduction and nurturing. Erich Neumann (1963)

and others see this as evidence of a widespread worship of the Magna Mater, or Great Mother.

Neumann argues that the matriarchal world precedes the patriarchal world, both in the development of individual consciousness and in the development of mythology. But he is not talking about societal structures. Patriarchal societies have predominated throughout the world from the earliest times of recorded history, and there is no clear reason to assume the positions of power were not held primarily by men in earlier millennia. Yet as infants we all begin life in a realm where the mother is central, and Neumann contends that worship of the Great Mother was once the primary focus of mythologies throughout the world.

Christine Downing (1981) notes that the clay female figurines date from a period when conscious agriculture emerged throughout Europe. Perhaps it seemed natural to our ancestors who depended so heavily on the uncertain production of food to think of the earth as like a mother who can bring forth life and provide nourishment. If you demonstrate your love for her, perhaps she will meet your needs. Widespread worship of the Great Mother, then, may have preceded the emergence and dominance of powerful male figures in Indo-European pantheons.

A mother figure has sometimes assumed a central role in the thinking of various groups in more recent, historical times. The first-century worship of Artemis in Ephesus is a case in point. It is also possible that peaceful agricultural societies are more likely to favor a Great Mother goddess than are hunting and warring societies, where the role of aggressive men is reflected in myths that focus on powerful male gods. It appears that mother goddesses predominated among the Canaanites and other people who occupied the region of the Levant before the Hebrews arrived and demanded worship of their male god. Worship of the goddess continued, but the growing role of Yahweh was consistent with what was happening in other parts of the Mediterranean world. There was a widespread tendency, whether demanded by societal authorities or not, to subordinate goddesses and accord greater attention to powerful male deities.

7. Mythic Modes of Action and Empowerment

We can now take a closer look at some specific gods, goddesses, and heroes of mythology. If we attempt to group them according to the human qualities they embody, it is evident that many of the obvious categories correspond to the qualities that appeared as factors in Chapter 3. In short, we are dealing with qualities that we commonly regard as either masculine or feminine.

For each quality considered essentially feminine, we find a preponderance of female figures in various mythologies. For each quality considered masculine, we find a preponderance of male figures. Yet in each case, it is possible to find both male and female figures who display the given quality.

I do not pretend to offer an exhaustive presentation of any one mythology. For our present purposes, it is more useful to focus on selected figures for whom available accounts offer a view of the personal traits they display. I also see no need to confine attention to figures drawn from traditional myths. In popular imagination, mythic qualities are often ascribed to figures drawn from popular folklore, fiction, history, and the contemporary public scene, and it is worth noting these wherever they fit.

NURTURANCE AND COMPASSION

I have already noted a class of goddesses who embody nurturance — the great mother goddesses who commanded attention at an early point in cultures spread across the globe. They include such figures as Gaia in Greece, Mut in Egypt, Cybele in Anatolia, and Aditi at one stage in Hindu mythology. Teutonic myths are best preserved in Norse writings, where we find Audhumbla and Frigg playing prominent roles as mother figures. Among the Pueblo Indians, where weaving has long been important, Spider Woman appears as a creator mother goddess.

In China, among Buddhists and Taoists, Kuan Yin is viewed as a divine mother goddess. She is seen as an embodiment of pure compassion. Like Artemis in Greece, she is often considered a virgin goddess who protects women. In many respects her role is like that of the Virgin Mary in the experience of many Christians in the Western world. Kuan Yin is revered in Japan and Korea as well as China. She can be traced historically to India, for her name is actually a translation of the Sanskrit Avalokitesvara, the merciful god of Indian and Tibetan Buddhism.

Demeter was another important mother goddess of Greece. She was also a goddess of vegetation, associated particularly with the production of grain. The best known story of Demeter concerns her relationship with Persephone, a daughter with whom she was closely bonded. It is said that young Persephone one day was out in a meadow picking flowers with other maidens when she spotted a beautiful narcissus. She ran to it and reached down to pluck it, when the earth suddenly opened up. Hades, the lord of the underworld, sprang forth with his chariot and his powerful steeds. He seized the girl and proceeded to carry her off to his realm. She shrieked, pleading for help from her father Zeus. Zeus remained comfortably on his throne on Olympus and did not hear her, but her mother and a few other immortals did hear her cry. Demeter felt an intense pain in her heart and did not understand what had happened to her beloved daughter.

For many days, Demeter wandered the earth grief-stricken, neglecting her care of the crops. She made inquiries of other immortals but did not learn the truth until she encountered Helios, who drove the chariot of the sun. He informed her that Zeus had given his brother Hades permission to claim Persephone as his bride. Inconsolable and angry toward Zeus, she continued wandering the earth, going from city to city and neglecting her appearance. Shrouded in grief, she soon manifested the aspect of a frail old woman. She came to the palace of a wise king named Keleos. The four daughters of Keleos saw her and asked how it happened that she had come to the palace. Demeter did not disclose her true identity but told a fanciful tale in which she had escaped from pirates. The young women invited her into the palace.

Demeter offered her services as a nursemaid, as a servant, and as one who could teach handicraft to other women of the household. It so happened that Keleos and his queen had a newborn son, and the infant was entrusted to the care of Demeter. She sat quiet before the queen, mourning her daughter, until a handmaiden managed to cheer her up and she was able to laugh and talk. The queen saw something regal in Demeter's appearance and entrusted the care of her young son to the goddess. Unbeknownst to the parents, Demeter would anoint the boy with ambrosia, and at night she held him up to the fire, exposing him to its full strength. In her care, he grew splendidly and would in time have become immortal.

Alas, one night the queen peeked in and saw what Demeter was doing. She screamed, believing the goddess was killing her son. Now angry, Demeter put the boy down and berated the queen. She then revealed her actual identity and resumed her true appearance as a goddess. Keleos had a temple and altar built to honor Demeter, and Demeter sat for a long time in the temple, mourning her daughter. For a year, she willed a dearth of crops upon the land. No seeds would sprout, and oxen pulled the ploughs through the fields in vain. Zeus took notice of the resulting devastation and sent a variety of gods and goddesses to plead with Demeter, but their efforts were futile. Demeter said that the earth would bear no more fruit until she saw her daughter once again. Then Zeus sent Hermes, the swift messenger god, to the underworld to ask that Hades send Persephone back into the light. Hades acceded to his brother's wishes and told the maiden that she could return with Hermes in his chariot. Before she left, he placed pomegranate seeds in her mouth, and she consumed them. She was soon reunited with her mother, to the great joy of both, but since she had eaten the pomegranate seeds, she was compelled to spend a third of every year in Hades as queen of the underworld.

In Egypt the goddess Mut was viewed at an early time as a great world mother who brought forth all living things. Yet as the mythology evolved, she was succeeded by Isis, who was seen for many centuries as the greatest goddess of all, a loving and beneficent mother who created, protected, and provided for all creatures. By the first century of the common era, the cult of Isis had spread beyond the borders of Egypt to neighboring areas around the Eastern Mediterranean region. She was worshiped by many Greeks and Romans as a goddess who could guarantee the growth of crops and protect sailors and their ships.

The best known story of Isis depicts her as a loving wife and mother. She and her husband Osiris were joined in marriage by the sun god Ra. They enjoyed an idyllic marriage and were admired by everyone, except for Osiris's brother Set. Set was jealous of Osiris and managed to slay him and conceal him in a coffin, which was then cast into the Nile and carried down to the sea. The grieving widow searched far and wide for her husband and managed at length to retrieve his body. But Set managed to steal the body, which he cut into fourteen pieces and scattered in various parts of Egypt.

Isis then turned herself into a bird and flew up and down the Nile in search of the pieces. She found all but one, and she managed to reassemble the body. Only the phallus was missing, and she formed one of gold and wax and set it in place. Prior to this, she and her husband had been childless, but with her magical skills, Isis was able to revive Osiris briefly and conceive a son. She reared her son Horus with loving care out of sight of Set. When Horus was fully grown, he was able to challenge Set, avenge the death of his father, and assume his own place on the throne of Osiris.

Another nurturing figure in Egyptian mythology is the Nile god Hap (also Hep, Hapi). He is depicted as a man with the breasts of a woman. This is consistent with his role as a god of fertility and nourishment, one who can ensure the growth of crops along the river.

There is a parallel to the story of Isis in the best known tale of Orpheus. Orpheus was not viewed as a nurturing god. Rather, he was the preeminent singer, who with his voice could enchant all humans, animals, and gods. He was able to save the Argonauts from disaster by overcoming the song of the sirens with his own. Orpheus thus stands out as a mythic figure associated with music, poetry, and the arts.

The present tale, however, depicts him as a loving and devoted husband, who grieved for the loss of his wife Eurydice. Eurydice died when bitten by a snake. Unable to withstand the loss, Orpheus managed to travel to the underworld to retrieve her. He pleaded with Hades and sang. Hades was so charmed by the song that he agreed to let Orpheus take his wife back with him, with one provision — that Eurydice was to follow her husband and he was not to look back until they had returned fully to the land of the living. Orpheus was overjoyed and began the journey back to the surface. Unfortunately, in his eagerness to be reunited with his wife, he momentarily forgot his instructions as he neared the end of the journey, and he turned to look at his beloved. At that instant, she disappeared into the darkness, and he lost her forever. Until his death he was unconsolable.

The story of Orpheus is not a story of general compassion — i.e., a tale that demonstrates a general concern for the distress of others. It is rather a story in which two individuals show a deep passion for each other. In this respect, it foreshadows the medieval romances that emerged in later centuries. Perhaps the best known of these is the story of Romeo and Juliet. The two young people in this tale fall in love, but unfortunately they come from two feuding families, the Montagues and the Capulets. They pursue their love and exchange wedding vows in secret, but it is not possible for them ever to live together in peace because of the bitter enmity that separates their families. In the end, each of the lovers chooses to die, believing that the other is already dead. Thus they both die trusting that they will be united forever in death.

A romance of comparable renown is that of Tristram and Isolde. (The names appear in various other forms — Tristan, Tristrem, and Tristran in the one case and Isolt, Iseult, and Yseult in the other.) In this story, Tristram is sent to Ireland to bring Isolde back to Cornwall to marry his uncle King Mark. On the return journey, however, both Tristram and Isolde unwittingly swallow a potion that causes them to fall madly in love. They pursue many trysts, which vary from one version of the story to another, but ultimately they become estranged. Isolde then marries Mark, and Tristram marries another woman. As Tristram lies dying of a battle wound, however, he sends for Isolde. He dies believing that she is not coming. She arrives too late, and finding him dead, she too dies in a state of grief. A tragic ending appears typical of medieval tales of romantic love.

Christian literature contains many accounts of people who manifested a more general compassion for others. Jesus of Nazareth was crucified in the early decades of the first century, but in gospels that appeared by the end of that century his life story had acquired many mythic features. In Chapter 6 of the gospel of Mark, there is a story that demonstrates his compassion. In a desert area he had managed to gather a crowd of five thousand people who had come to learn his teachings. Toward the end of the day, his disciples came to him and suggested that he send the people back to the villages so they could buy bread for their evening meal. But Jesus instead asked how many loaves he and his disciples had. Told that they had but five loaves and two fishes, he asked that all the people gathered there be seated on the grass. Then he managed miraculously to multiply the loaves and fishes, enabling everyone to be fed.

Many other figures in Christian history have been regarded as exceptionally compassionate. Jesus's mother Mary is so viewed, and in the minds of many people she has acquired the de facto status of a goddess. A Christian saint of whose life we know much more is Saint Francis of Assisi. He lived from 1181 (possibly 1182) to 1226. Born the son of a wealthy merchant, he served as a soldier and was a prisoner of war for one year. He could have gone on to live a life of luxury, but in his early twenties he had a conversion experience. He could no longer pursue the life of a soldier nor that of a merchant. He exchanged his costly garments for the cloak of a beggar and gave away his earthly goods to the poor. He and the young men who had joined him in his mission went about Umbria, humbly attired, aiding people in any way they could. He went to see the pope, who recognized him and his followers as members of a new order within the church. Many poor people who had grown disillusioned by church practices were drawn back into the fold by the work of Francis, and the pope sent him on diplomatic missions to several other countries.

Another saint known for his benevolence was Saint Nicholas, patron of children and sailors, who was evidently a bishop in Asia Minor. Few solid facts are known about his life, but he is the subject of many legends. It is said that he saved the daughters of a poor man from a life of prostitution by throwing a bag of gold through their window on each of three successive nights, thereby providing them with dowries that enabled them to secure prosperous husbands. He is also credited with restoring life to various people, including three little boys who had been chopped up and salted by a butcher with the intention of serving them as meat for his guests. It is said that he made the sign of the cross over the tub that held them, and they then stood up, fully revived. In colonial America, English colonists learned of Saint Nicholas from the Dutch and rendered the name they heard (Sint Nikolaas) as Santa Claus. Needless to say, Saint Nicholas, or Santa Claus, has since become the subject of many fanciful tales — most notably "A Visit from Saint Nicholas" by Clement C. Moore. Throughout these tales he remains a legendary figure known for providing gifts for children.

It is possible to cite many exceptional cases of altuism in more recent times. An example is Raoul Wallenberg, a Swedish diplomat who is credited with saving 100,000 Hungarian Jews from extermination by the Nazis during the second world war. At great risk to his own life, he issued Swedish passports to many of them and found ways to shelter and hide many others in houses. Mother Teresa, who died in 1997, is another example. At 17, she went to India, where she became a Roman Catholic nun and taught school in Calcutta. Ultimately she left the convent so she could devote her life fully to helping the poor. The Missionaries of Charity, an organization that she founded, operates schools, hospitals, orphanages, and food centers in more than ninety countries.

Nurturing maternal figures abound in folk tales. In myths of great heroes and in folk tales depicting the lives of undistinguished mortals, we often find a developmental sequence in which a nurturing mother is followed by a devouring mother. The latter may take the form of a dragon or some other monstrous beast. In the tale of Hänsel and Gretel, as recounted by the Grimms, the two children have been abandoned in the forest, and they are drawn to a cottage made of bread, cake, and sugar candy. Since they are very hungry, they proceed to nibble at the dwelling, but the cottage is the home of an evil witch who has lured many children there, and she intends to bake them and gobble them up. They must overcome her before they can return home. The sequence from nurturing mother to devouring mother corresponds most clearly to a stage in early childhood when children experience an urge to break free from the one-ness of the mother-infant bond and assert their individual separateness and emerging self-awareness. We see an expression of this urge in the negativism of the two-year-old.

It is not too surprising, then, that the devourer is sometimes the dark side of a nurturing mythic figure. Yet there are times throughout life when each of us needs the aid of someone who can nurture, and there are many times when we may feel a desire to nurture others. In some of the relationships with the most important people in our lives, we may take turns playing a nurturing role.

I did not choose a profession that requires me to play a nurturing role, but I have often been aware of an urge to nurture. I have felt close to my children since they were infants, and I value the time I spent caring for them and teaching them things in their early years. I also love small animals and enjoy the four cats who share our household. I feel drawn to the small animals I see in the desert areas near my home, and I have fond memories of the morning when I rescued a baby javelina and fed it water with an eyedropper. It had become separated from the rest of its family and left stranded in the shrubbery of a neighbor's front yard.

Do you like to play a nurturing role? Are there benevolent beings that we noted above with whom you feel you have something in common? Are there nurturing or compassionate people with whom you identify? To what extent do you regard nurturance or compassion as part of your own nature? Are these qualities you feel a desire to cultivate?

ASCENDANCE

In every major mythology we can find accounts of gods who ruled and gods who sought power. In Greek mythology a major sequence begins with Uranus (or Ouranos), the god of the sky. Each night Uranus came down to mate with the earth goddess Gaia. As a result, Gaia bore six sons and six daughters, but as soon as each was born, Uranus hid it within the earth and would not let it come into the light. Gaia suffered from this inner burden and finally assembled her children to ask for their aid in punishing their father for the oppression he had imposed on them. Only the youngest son, Cronus (Kronos) had the courage to volunteer. Gaia equipped him with a sickle, and that night when Uranus came to mate with Gaia, the son emasculated his father.

Cronus then became the most powerful god. He married his sister Rhea, and she bore three daughters (Hera, Demeter, and Hestia) and three sons (Hades, Poseidon, and Zeus). He was already more powerful than his brothers, but he had been forewarned that he was destined to be overthrown by one of his sons. Not willing to be outshone by anyone, being the glorious king that he was, he swallowed each of his children as soon as it emerged from Rhea's womb. After losing five children in this manner, grief-stricken Rhea consulted her parents and then took their advice. Pregnant with her sixth child, she slipped away to the island of Crete, gave birth to Zeus and left him in a cave in the care of Gaia. She returned to Cronus with a stone wrapped in swaddling clothes, and he immediately swallowed it.

Zeus quickly grew to manhood and overpowered his father, who was forced to disgorge the rest of his children. As supreme ruler, Zeus was not a tyrant like his father and grandfather. He established a just world order, punishing transgressors but aiding those whose causes were right. He shared some power with his brothers, giving the realm of the sea to Poseidon and the underworld to Hades, while remaining in charge of the sky himself. He is the subject of innumerable tales. In most, he is married to his sister Hera, but he has many erotic adventures with other feminine figures, both mortal and divine, leading to some resentment on the part of his wife.

We can find a succession of supreme gods in other mythologies. In Egypt at one time, we find such gods as Amun and Ra, the sun god, in a ruling position. Sometimes those two were viewed as a single god, merging as Amun-Ra. Then Seb, the son of Ra, appeared and was said in some hieroglyphic inscriptions to be the father of the gods. It was Seb's son Osiris, however, who came to be worshiped throughout Egypt, alongside his wife Isis. Like Zeus, Osiris was seen as a just ruler and as a personification of moral virtue and truth. His brother Set, who slew him, was viewed as a personification of evil.

In Teutonic mythology the supreme god was Odin (or Woden, Wotan, and other variants of the name). He occupied a position comparable to that of Zeus and shared power with his wife Frigg, who exercised control over nature. As we turn to India, we find a variety of gods described in the earliest known Hindu writings, the Vedas. These include Dyaus, Varuna, Indra, Surya, and Savitar, among others. In later writings we find a trinity of deities — Brahma, Vishnu, and Siva (or Shiva). These three may be roughly categorized as the creator, the preserver, and the destroyer, and they are actually viewed as three aspects of one entity, which we might characterize as the divine essence, or the ultimate ground of being and consciousness. This essence, as an ultimate unity, is neither masculine nor feminine. It lies beyond all gender distinctions, as well as distinctions of any other kind. It is commonly referred to as Brahman.

Ruling figures are not always male in early myths and legends. In early writings of the Greeks — e.g., Homer, Aeschylus, and Herodotus — we find accounts of Amazons, a society or tribe of nomadic women warriors, ruled by a queen (Hippolyte in one account). They used men only to become pregnant, though they sometimes came to the aid of men. According to Homer's *Iliad*, they aided the Trojans in their battle with the Greeks. Jeannine Davis-Kimball (2002) believes that stories of the Amazons may have been inspired in part by actual warrior women, though those women may have fought alongside their men rather than maintaining an exclusively female society. Archeological findings in the southern Russian steppes may contain relevant evidence, and Davis-Kimball believes that descendants of the ancient warrior women may be found among present-day nomadic people in that region.

As we turn from ancient myths to medieval legends, the most enduring figure we find is King Arthur. The early accounts of him are rooted in Celtic legends of the Irish, Welsh, and Cornish people, but some were probably inspired by actual men who fought against Saxon and Roman invaders of Britain. As the stories flourished, they were spread around the European continent by minstrels from Brittany, and they were later embellished further by writers like Malory and Tennyson. According to the most basic versions of his life story, Arthur was the illegitimate son of King Uther Pendragon. Reared in secret, he demontrated his true identity as successor to the king when he drew the sword from the stone and the magician Merlin revealed his parentage. He proved to be a noble king, and he assembled a group of knights and seated them about the round table. This physical arrangement symbolized the equality of all these men, and they were all expected to perform good and noble deeds. Upon his death, Arthur was borne away on a boat tended by ladies from the isle of Avalon, and it is expected that he will return one day in Britain's hour of greatest need.

Many historical leaders have acquired mythic qualities in the minds of many people and have been both idolized and demonized. Obvious examples are Julius Caesar, Charlemagne, and Russia's Peter the Great. The same can be said of some of our presidents, notably George Washington and Abraham Lincoln. It is more difficult to think of recent presidents from a comparable distance. We are too aware of their flaws. (Nonetheless, both John Kennedy and Ronald Reagan have their fans.) In any case, you can probably think of noteworthy people, male or female and past or present, who represent your own ideal of leadership.

We have now noted a number of mythic and historical figures who have been in positions of power. They have been directly in command of other people or have been in a position to make decisions that affect many people. You probably admire some people who are or have been in positions of power. Would you like to be in such a position yourself? Personally I have never sought a position of power. I do not enjoy being in charge of other people, though I have enjoyed occasional situations in which I have directed assistants on research projects. In interacting with another adult, I usually prefer to play a cooperative role, rather than a dominant or submissive one. Of course, there are occasional situations where it makes sense for one person to take charge — e.g., when one of us is showing the other how to do something. I am often disturbed by the decisions made by leaders in political office, but I have never felt tempted to run for an office that might enable me to have a greater influence on those decisions. Perhaps there are things I could accomplish if I sought directive or administrative roles.

What about you? Have you sought or assumed dominant roles? Would you like to do so? Are there people or mythic figures we have noted above with whom you feel you have something in common? Are there leaders of the past or present with whom you identify? To what extent do you regard dominance or ascendance as a part of your own nature? Is this a quality you feel a desire to cultivate?

AGGRESSION

Deities and people in positions of power may employ aggression to maintain or extend their rule. There are also those who relish aggression for its own sake. In mythology, many of them are known as gods or goddesses of war. A case in point in Greek mythology is the god Ares. He represents the very ferocity of fighting, and his father Zeus deemed him the most hateful of the gods.

In Greek mythology we find other figures, such as the Amazons, simply known for engaging in battle. There are others, like Artemis, who occasionally resort to violence to mete out punishment. Artemis, goddess of the hunt, was a protector of women and young animals, but when the young hunter Actaeon came upon her unexpectedly as she was bathing, she responded by turning him into a stag. As he then fled, he was ripped to pieces by his own hounds.

In Teutonic mythology, the equivalent of Ares would be Tyr (or Tiw), the god of war and athletic contests. Yet he is more honorable than Ares, for he is known for courage and adherence to law. He can be trusted to play by the rules and keep his word.

The Morrigan, a Celtic war goddess, was more comparable to Ares. She hovered over battlefields, incited soldiers to fight and relished destruction. A contrasting figure of Celtic myth was Cuchulain, the supreme warrior of Irish folklore. He displayed unusual strength and athletic skill early in childhood and became a warrior before he was fully grown. He easily defeated powerful enemies in battle and defended Ulster, at times single-handed, against an invading army.

In Hindu mythology, of course, Shiva represents the destructive aspect of Brahman. And the goddess Kali, sometimes venerated as a divine mother, also appears as a destructive and terrifying figure, garlanded with skulls and holding a bloody sword. Hindu deities, of course, can manifest themselves in more than one form, and it is possible to view these two forms of the goddess as either alternative aspects of one goddess or as two separate goddesses.

In Egyptian mythology too, we find deities who manifest in more than one way, commonly with different names. Thus the goddess Hathor, a protectress of women, becomes Sekhmet as a goddess of war and a merciless slayer of men. Hurling herself against men who had rebelled against Ra, she enjoyed slaughtering them and drinking their blood, and Ra had to contrive a way of stopping her before she could destroy the entire human race.

Moving closer to the contemporary world, we can find many military figures who have been idolized. And in the realm of athletics and sports, there are many athletes and players who have gathered fans. These would include wrestlers, boxers, masters of the martial arts, and the heroes of football, hockey, baseball, and other sports. In some sports, there are outstanding women as well as men.

Movies and television have provided an additional array of fictional characters known for aggression, strength, or fighting ability. On the screen we encounter such male figures as the Incredible Hulk, Rambo, and the superheroes Superman and Spider Man. Movie Westerns like *Shane* and *High Noon* typically contain a hero who saves the town or the valley by killing the bad guys before he rides away into the sunset. The Amazons are represented on TV by such female heroines as Wonder Woman and Xena, the warrior princess.

I personally abhor violence and have no desire to watch people beating each other up. I believe there is much unnecessary violence in the world, and I consider it irresponsible for any government to employ military aggression as a primary mode of conducting foreign relations. I believe violence usually creates more problems than it solves. There have been times, however, when I felt that my rights or those of someone else had been violated. On those occasions I responded very assertively or vigorously pursued a course of action that I hoped would correct the situation. As for sports and athletic events, I enjoy demonstrations of athletic skill. For me, that is the most interesting part of any competitive team sport, and I have never cared which team won any game. I prefer watching gymnastic and track and field events. I realize that some people regard team sports as a useful way of expressing their aggressive and competitive impulses, whether they are participants or spectators. Do you? Obviously I do not expect your tastes to be the same as mine with respect to sports, physical activities, or movie and TV dramas.

Are there people, mythic beings, or fictional characters we have noted above with whom you think you have something in common? Are there aggressive, assertive, or competitive people with whom you identify? To what extent do you regard aggressiveness, assertiveness, or active competition as a part of your own nature? Do you have a desire to cultivate such a quality?

AUTONOMY

In mythology we find a variety of gods and goddesses who operate mostly in solitude, avoiding close relationships. One such god was Hephaestus, the lone artisan and inventor of Olympus. Both his parents, Zeus and Hera, rejected him. He was married to Aphrodite, but she largely ignored him and pursued an affair with Ares. He built palaces for other deities and devised weapons for many of them. Working mostly at his forge, he created arrows, armor, thunderbolts, necklaces, and a winged chariot. He devised two beautiful golden maidservants who could speak and obey his requests. Since he was lame, they would hasten to his side to assist him when he walked. It is said that he also created the first woman, Pandora, at the request of Zeus.

On occasion he devised traps for those who had wronged him. He once sent Hera a magnificent golden throne. She was delighted when she saw it, but when she sat on it, she was immediately seized by invisible hands, and only Hephaestus was able to free her. He also created an invisible net that confined Ares and Aphrodite when they ventured to go to bed together. He displayed the trapped couple before other Olympians, hoping they would share his anger at being thus cuckolded, but the scene merely evoked a great chorus of laughter. Ceaselessly working, he created many ingenious devices and works of art. Hephaestus could be considered the prototype for many creative scientists, inventors, and artists.

The Greek goddess Artemis was also a very independent deity. Immediately after her birth, she asked her father Zeus for a short tunic, hunting boots, and a bow and quiver full of arrows. As a huntress she roamed the hills and valleys with her hounds, but she was often accompanied by a group of nymphs. She vowed to remain a virgin and resolved to avoid becoming attached to any man. She became enamored of the young hunter Orion, however, but unintentionally killed him. Orion had swum far out in the sea one day, and her brother Apollo challenged her to shoot an arrow that would stike an object that they saw bobbing up and down on the horizon. Accepting the challenge, she struck Orion in the head. At times, Artemis was known to kill both men and women when they aroused her wrath, but she could also be protective of people in need.

The goddess Athena also remained chaste and was evidently devoid of any erotic attraction to the males who found her of interest. According to one story, she was born from the head of Zeus. Zeus had swallowed the pregnant goddess Metis and began thereafter to have an intolerable headache. To relieve his pain, Hephaestus opened his skull with a bronze axe, and Athena burst forth fully armed and uttering a loud shriek of victory. Athena came to be Zeus's favorite child, and she became an ardent supporter of his patriarchal rule. In some accounts she played the role of a warrior goddess, but she could also be a peacemaker. In some tales, she came to the aid of heroes in combat. According to Homer's *Odyssey*, she came to the aid of both Odysseus, as he sought to return home from Troy, and his son Telemachus. She was also known as a goddess of arts and crafts and as a goddess of wisdom.

Hephaestus, Artemis, and Athena are strikingly different from one another, but they share a determined independence. People in various walks of life relish a sense of self-sufficiency. They may value the freedom for creative work that living alone provides, or they may simply pride themselves on being able alone to meet all their needs. Daniel Defoe's novel *Robinson Crusoe* depicts a shipwrecked mariner who succeeds in developing a repertoire of survival skills while stranded alone for many years. This theme was revived in the movie *Cast Away*, starring Tom Hanks.

I said earlier that I did not like to be in positions of power, directing other people. On the other hand, I do not like positions in which I am repeatedly directed or ordered about by someone else. I prefer work in which I can follow my own bent and be my own boss. I enjoy interacting with other people, but I dread days in which I am constantly interacting. I would rather have some time alone. During the periods in my life when I have lived alone, I have been aware that my solitary status gave me the freedom to do creative things without having to mesh my schedule with that of anyone else. Yet when living alone, I have always been aware of a desire to have a close relationship and to share significant moments with someone else. I look for a balance, but there are people who prefer a permanently solitary life, and there are people who prefer constant interaction with other people. What is your preference?

Are there independent people or beings with whom you feel you have something in common? Can you think of any such people you know or people noted for their autonomy with whom you identify? To what extent do you regard autonomy as part of your own nature? Is this a quality you feel a desire to cultivate?

8. Mythic Modes of Consciousness

We turn now to gods, goddesses, and heroes noted for various distinctive modes of thinking, feeling, and imagination. We may consider some of these modes masculine and others feminine, but it is possible for each of them to find both male and female representatives whether we search in mythology or in human history.

ORDERED, RATIONAL THOUGHT

As the ruler on Mount Olympus, Zeus was called upon to make wise, rational judgements and decisions. But rational judgment may be a more distinctive quality of Apollo. Apollo was the son of Zeus and Leto and the twin brother of Artemis. He is the subject of many stories, and in them he plays many roles. He was an athlete who displayed incredible strength and skill as an infant. He was a superb archer. He played the lyre and was considered a god of music. He protected shepherds and their flocks. He was known also as a god of divination and prophecy.

Apollo had a varied repertoire, but in each of his roles, he tended to maintain an emotional detachment and weigh his choices before proceeding, avoiding impulsive actions. He liked order and harmony. If he were around today and involved in music, he would prefer Bach to bebop or jazz improvisation. He was a giver and interpreter of laws, and he provided communities with the legal structures they needed to resolve disputes fairly.

A goddess known for rational judgment, who appeared early in Greek mythology, was Themis. When Zeus began his reign, he chose her for a wife. When he later married Hera, Themis remained by his side to offer advice and counsel, and Hera, though noted for her jealousy, accepted Themis in that role. On Olympus, Themis sought to maintain order, and it was she who oversaw ceremonials and brought the other Olympians together for such occasions. If she were assisting in preparing for a dinner party today, she would be the one who would write the names of guests on cards and put them at the appropriate place settings. On earth she served as a goddess of justice. Like Apollo, she ensured that the innocent would be protected and the guilty punished.

Another Greek goddess known for her wisdom was Athena. She has received much more attention than Themis, and like Apollo, she is the subject of many stories. She is considered the daughter of Zeus and Metis, the first wife of Zeus. According to one story, Zeus was forewarned that any children he had with Metis might be more powerful than he and dethrone him. To forestall such a fate, he swallowed Metis when she became pregnant. As a result, he experienced terrific headaches. To relieve his pain, Hephaestus opened his skull with a bronze axe, and Athena burst forth fully armed. Zeus came to love her as his favorite child. Yet the circumstance of her birth demonstrated her propensity as a warrior goddess. She was indeed a warrior goddess but was not known so much for taking part in battle as for protecting certain warriors and for taking part in the planning for battle. In Homer's *Iliad* she provided tactics for the Greeks in their war with the Trojans. In the *Odyssey*, she aided Odysseus in his efforts to return home to Ithaca and helped him become reunited with his wife and son.

As a virgin goddess, Athena avoided close relationships and maintained an emotional distance, preferring like Apollo to act on the basis of sound rational judgement. She was known as a goddess concerned with practical skills and crafts, particularly weaving. Such a skill requires careful planning and patient, diligent work in the execution of the plan.

Perhaps rationality and order are best represented in the Egyptian pantheon by the god Thoth. He intervened in the battle between Horus and Set, who had slain Osiris. He healed the wounds of both, and then, before a tribunal of gods, he resolved their dispute, establishing Horus's right to his father's throne. Thus he was a wielder of justice, but above all, he was the divine scribe, record-keeper, and scholar. It was he who invented all the arts and sciences: arithmetic, geometry, astronomy, medicine, surgery, music, drawing, writing in hieroglyphics, magic, and soothsaying. He was the keeper of the divine archives and patron of history, science, and literature.

Two goddesses, Seshat and Maat, were considered wives of Thoth, and each of them shared some of his functions. Seshat was a goddess of history and, along with Thoth, a record-keeper for the gods. She was also involved in the invention of writing and was the goddess who measured time. As mistress of the house of architects, she helped establish the proper orientation and design of temples. Maat, on the other hand, was associated more with Thoth's judicial functions. She was considered the goddess of law, truth, and justice.

The closest approximation to a god of wisdom in Hindu mythology is the god Ganesa. A popular divinity, he is considered calm, friendly, and lovable and usually depicted with a fat belly and the head of an elephant. Combining the wisdom of both elephants and people, he is deemed a god of prudence and policy. Furthermore, he is a god of literature, and his name is commonly invoked at the opening of Hindu literary works.

Orderly, rational thought has been ascribed to many historical figures, some of whom have acquired a mythical aura in popular imagination. The biblical figure who most clearly embodies rational wisdom is King Solomon. In the realm of mathematics, logic, and science, we might note such people as Euclid, Aristotle, Sir Isaac Newton, and Albert Einstein. Among the leading founders of our republic, Thomas Jefferson, John Adams, and Benjamin Franklin are commonly regarded as exemplars of rational thought. We can find many examples in fiction. Sherlock Holmes is an obvious case in point.

Women as well as men have excelled in fields calling for superior rational thinking when the time and setting have provided the opportunities for them. Marie Curie, a chemist and physicist, worked for a time on uranium and reported a probable new element in pitchblende. Then, she was joined by her husband Pierre, and the two together discovered polonium and radium. Lise Meitner, an outstanding figure in the field of atomic physics, made major contributions to our understanding of nuclear fission. Hildegard von Bingen was one of the great intellectual figures of the twelfth century. As a German nun, she became known as a composer, writer of lyrical poetry, and scholar. She corresponded with many of the leading political and intellectual people of her day. Her writings include a medical encyclopedia, works of natural history, scientific treatises, and treatises on the lives of saints.

Personally I cannot claim to have the wisdom of both elephants and people. On the other hand, I have had a career as a professor of psychology, and it is hardly possible to prepare academic lectures and write scholarly books without considerable use of ordered, rational thinking. Do your chosen activities or your job make a similar demand? Orderly, rational thought is a quality important in mathematics, science, law, diplomacy, business planning, and many other fields. Of course, you may find that you rely heavily on rational thought and orderly planning either more or less than most people in the ordinary course of living.

Are there people, mythical beings, or literary characters we have noted above with whom you think you have something in common? Are there rational, systematic thinkers with whom you identify? To what extent do you regard rational thought as a part of your own nature? Do you have a desire to cultivate such a quality?

Intuitive Wisdom

Rational thinking provides an avenue to truth, but it is not the only possible avenue to truth, and the truth to which it leads may not be the only kind of truth. Rational thought proceeds by careful examination and analysis of available content, whether in print or in the world about us. Intuition proceeds by allowing images, ideas, and meanings to emerge. It is the key to extrasensory perception, to artistic inspiration, and to the appreciation of symbolic meaning.

In considering a myth — say the story of Adam and Eve in the garden of Eden, as told in Genesis — the individual who relies exclusively on rational thought may take the story literally. For such a person, the story is either a factual, historical account of the first humans or it is patently false and not worthy of further consideration. Someone relying more heavily on intuition would be less concerned with the literal meaning and sense a symbolic meaning having to do with our emergence long ago as a distinct species and with our emergence in infancy as individual conscious beings.

In Greek mythology, the man who most clearly embodies intuition is Tiresias, a seer of many legends. In the course of his effort to return home, Odysseus consulted the spirit of Tiresias in the underworld. Tiresias was blind, and there are several stories that account for his blindness and for his special gift. According to one, he was blinded by Athena when he caught sight of her bathing, but in compensation she granted him the power of divination. This combination of attributes turns up in stories of other seers, the implication being that a person deprived of outer sight may balance that loss through development of inner sight, or "insight."

One of the unusual features of Tiresias's life is that he spent seven years as a woman. We are told that he was turned into a woman when he wounded a couple of copulating snakes on Mt. Cyllene and restored to manhood when he encountered them again at a later time. As a result of these transformations, he possessed a kind of knowledge that the rest of us may wish for in vain. As a consequence, he was consulted to settle a dispute between Zeus and Hera. Hera complained of her husband's many sexual adventures with other women and goddesses, and Zeus dismissed the complaint, saying that it is well known that women derive more pleasure from sexual intercourse than men do. On this occasion, Tiresias sided with Zeus, saying that "the woman's pleasure is as ten to the man's one."

If we look for female figures in Greek mythology comparable to Tiresias as an embodiment of intuitive wisdom, we do not find anyone with such distinctive personal qualities. The goddess Sophia certainly represented a kind of intuitive wisdom, but she was seen a kind of primordial female principle and never described in personal terms. She plays a prominent role in the writings of Gnostic Christians in the first and second centuries.

There were a number of oracles, or oracular shrines, in ancient Greece and surrounding countries, and divination was usually the role of a priestess. It was she who received the messages from the gods, perhaps after she had entered a trance, but often a priest was called upon to interpret the words that she uttered. If we liken the role of the priestess to that of psychics in more recent times, it is worth noting that people who become known as psychics may be either men or women but are more often women.

Seers comparable to Tiresias appear in other bodies of folklore. Merlin, who figures prominently in the legends of King Arthur, is a clear example. Gandalf, too, is an outstanding fictional character of this type. He has gained popularity through the writings of J. R. R. Tolkien.

Siddhartha Gautama was the son of a king and grew up in a life of luxury, sheltered from any sight of the miseries of the world about him. He married and fathered a son, but at 29 he rode forth in a carriage and saw, on successive occasions, a sick man, a very old man, a corpse, and a mendicant monk. From these encounters, he realized not only the inevitability of suffering and death but also the serenity of the life for which he was destined. He decided to leave his comfortable home and seek greater understanding. He explored a number of spiritual paths, including yogic meditation and asceticism. After six years of wandering, he sat under a pipal tree, vowing to remain there until he had attained supreme enlightenment. The moment of illumination came after many days. At that moment, he became the buddha (one who has awakened). He formulated the basic doctrines of the four noble truths and the eightfold path and devoted the remaining years of his life, from age 35 to age 80, to wandering about and offering his teachings to others. Buddhism proceeded to spread throughout Asia, and in recent years it has gained increasing attention in the West.

With respect to its impact on world thought, the moment when the buddha achieved enlightenment may well be the most significant moment in the consciousness of any one individual. If there is a comparable moment that has affected thought in the Western world, it is the moment of St. Paul's conversion on the road to Damascus. Following the crucifixion of Jesus, he had undertaken the task of arresting Jesus's followers. Having arrested many in Jerusalem, he headed toward Damascus to continue his work, but according to the book of Acts of the Apostles, on the road he saw a blinding light, fell to the ground, and heard a voice that he understood to be that of Jesus. I have offered my understanding of that moment in my historical novel, *Shaul of Tarsos* (2004). That moment was the beginning of a lifelong mission in which Paul became the man primarily responsible for formulating the basic doctrines of Christianity that would appeal to people in the cultures of Greece and Rome and the man primarily responsible for spreading those doctrines beyond Judea. Without him, Christianity might have remained a minor Jewish sect. As a brilliant man, who was both Jew and Roman citizen and fluent in demotic Greek, he was uniquely qualified to do what he did.

We could cite other major religious figures, such as St. Francis and St. Clare, as exemplars of the kind of spiritual illumination we see in the life of Paul. We also find something comparable in the lives of psychic readers, notably in the life of Edgar Cayce, the most famous psychic of all. As a child, he claimed to have contact with the spirit of his grandfather, following the death of that man. On two occasions in his youth, he was able in a trance state to prescribe cures for physical conditions from which he suffered. There followed a lifelong practice of providing psychic readings for clients. Over 10,000 of his readings were preserved in written form, readings on health, past lives, ancient mysteries, and predictions for the future. He is alleged to have displayed an understanding of medicine and medical cures that he could not have gained through any formal training or systematic study. He is said to have shown his ability for precognition, retrocognition, clairvoyance, and telepathy, and he arrived at many of his insights in a trance state. He has been called the "sleeping prophet."

It is tempting to regard systematic, rational thinking and openness to intuition or inspiration as opposite and incompatible modes of cognition. Yet they are certainly not incompatible. One cannot do both at the same instant, but both are essential for most creative work. To produce something refreshingly new in art, in music, in poetry, in scientific theory, or in structural design, one must let go of existing order, be open to unexpected and unplanned ideas or images, and then work to develop new order. One must alternate between the two modes of cognition. I have always felt an urge to do something creative — whether in psychology, music, poetry, or fiction — and I am quite aware of this interplay. Yet I have known academic colleagues who are either so consistently disorganized or so intent on maintaining systematic order constantly that they never manage to do anything creative. People's tastes vary. We do not all have to have the same preferred life pattern or style of thinking. How would you describe your own?

Are there people or beings we have noted above with whom you feel you have something in common? Can you think of people you know or people noted for intuitive gifts with whom you identify? Do you have moments of sudden insight, sudden ideas or images, premonitions, or inspiration? Do you have dreams that offer you significant insights? Do you sometimes have a sense of being in communication with someone who is not in your immediate presence? Do you feel you are sometimes aware of events occurring at a distance or events that will occur? To what extent do you consider an intuitive gift, in any form, as a part of your own nature? Is it a quality you have a desire to cultivate?

AESTHETIC AND IMAGINAL PURSUITS

As I noted above, if we are open to the unexpected idea or image, artistic inspiration is one possible result. Of course, the moment of inspiration is just one part of aesthetic involvement. It is the initial step in the creative endeavor. The idea or image must be developed and executed in some form. One may be involved in the aesthetic realm as an imaginative creator, as a performer, or as a viewer or listener. We could say something similar about the work of the inventor, for invention too can spring from a moment of inspiration, followed by development of the initial idea or image, but the ultimate product may not be considered a work of art.

The Greek god Apollo, a god of music, was certainly involved in the aesthetic realm. Not only did he play the lyre, he was closely associated with the nine muses and often accompanied by them. The muses were patron goddesses and sources of inspiration and expression. Most of them focused on the arts and were gifted creators and performers in their own right. One muse was Calliope, the muse of epic poetry and eloquence. Enterpe was the muse of music, flute-playing in particular, and lyric poetry. Thalia was the muse of comedy. Polyhymnia was a muse of sacred and heroic poetry and later a muse of mimic art. Erato was a muse of love poetry. Melpomene was the muse of tragedy. Terpsichore was the muse of choral song and dance. Clio was the muse of history. And Urania was the muse of astronomy.

When Apollo played his lyre in the woods, the beasts would come from afar to listen. Yet there was a hero I have already noted — Orpheus — who is better known for his ability with song to enchant people, animals, and gods. Voyaging with the argonauts, Orpheus performed a number of miracles with his voice. He moved the ship from the beach to the sea, moved the terrible rocks that threatened to crush the ship, and lulled to sleep the dragon that guarded the golden fleece.

There was one miracle he could not quite accomplish — retrieving his dead wife Eurydice from the underworld. Yet we can construe his very ability to descend to the underworld and enchant the immortals there as the ability of a great artist to plumb the depths of his psyche, the sources of inspiration that lie far below the level of ordinary consciousness. His inability to possess what he finds there speaks to the tragic side of the artist's life. According to one of the accounts of the subsequent life of Orpheus, he was so inconsolable after losing Eurydice that he was torn apart by Thracian women when he failed to succumb to their charms. It is said that his head and lyre were then thrown into the river, where they continued to produce beautiful songs as they floated downstream.

Various gods and goddesses in other mythologies are associated with music. In the Egyptian pantheon, Hathor is seen as a goddess of love, fertility, and festivity. As one aspect of this role, she serves as a goddess of music and dance. She is often shown carrying a sistrum, an ancient musical instrument. The goddess Bast is also associated with festivals, music, and dance.

The muses of Greek mythology were daughters of Zeus. His counterpart in Teutonic myths was Odin. Odin was, among other things, a god of poetry, and his daughter Saga, as goddess of poetry, was more specifically identified with that art. In Scandinavian legends, Bragi appears as a god of poetry and eloquence. He was especially known for wit and cunning speech. In Hindu mythology, Sarasvati is considered a goddess of music, wisdom, and knowledge. It is said that she invented the alphabet for Sanskrit, and she is often shown holding a stringed instrument known as the vina.

A variety of gods and goddesses are regarded as creators of specific tools and instruments, but the one god who stands out as an inventor is the Greek god Hephaestus, who devoted virtually all his time to that pursuit. I have noted some of his creations in the previous chapter. Some of them were artistic, some were utilitarian, and some were designed to deliver punishment to those who had brought him grief.

Hephaestus might be considered the prototype for many inventors past and present. Thomas Edison and Nikola Tesla come to mind as outstanding historical examples. Following them in more recent years are the various people involved in the development of the computer and other electronic devices.

In the realm of visual art, the leading figures of the Renaissance come to mind as exemplars of artistic inspiration, in particular Michelangelo and Leonardo da Vinci. Many people would consider Michelangelo the greater artist of the two, but Leonardo's versatility is unique. He was a painter, sculptor, architect, scientist, engineer, and inventor. He was responsible for a number of architectural and engineering projects, but his notebooks contain plans for inventions, such as a flying machine and military weapons, that were never produced in his lifetime.

We might also note outstanding creators in other artistic fields. In music, Johann Sebastian Bach, Wolfgang Amadeus Mozart, and Ludwig van Beethoven come to mind. In literature, we might note William Shakespeare, Johann Goethe, and Dante Alighieri.

The aesthetic realm has always had a central place in my own life. Since childhood, I have been involved in musical composition. I have also enjoyed writing poetry and fiction. As a child, I studied piano and composition. At later points, I briefly studied violin, flute, and voice. I relished retirement from the university so I could devote more of my time to the arts. Of course, this is all a matter of preferred activities. What are your own preferences?

Are there people or beings we have noted in this section with whom you feel you have something in common (even if you do not have their extreme talent)? Can you think of people you have known who are involved in the arts or invention or such people who are so noted with whom you identify? Do you feel an urge to invent things or to create artistic works. Is enjoyment of visual art, music, dance, or literature important to you? To what extent do you consider the aesthetic or creative realm a part of your nature? Do you have a desire to cultivate such a quality?

PIETY

In Greek mythology the deity who most clearly represents piety is Hestia, the goddess of the hearth. To understand why tending the hearth might be considered a sacred duty, we must recognize that fire itself occupied a special place in the imagination of people in ancient times. It seemed to be a mysterious element that could readily appear and then vanish. It was valued for warmth and the preparation of food, but it was difficult to come by. A member of the family who was moving to another dwelling would take a piece of burning wood to the new home.

When towns formed, they commonly maintained a communal hearth from which families could obtain bits of fire. Furthermore, fire played an important role in ritual sacrifices.

There are few stories about Hestia, for she was largely confined to her sacred duty of tending the hearth. In pursuing this sacred duty, she declined offers of marriage from both Poseidon and Apollo and remained a virgin. In the maintenance of both household and communal fires, her role was often solitary, but in performing it she was seen at the same time as one who protected the home, the family, and the city as well.

In Roman mythology, the goddess Vesta played essentially the same role as Hestia, and the Romans regarded the two as the same goddess. In Rome a sacred fire was maintained and used for sacrifices. It was tended by a group of vestal virgins, originally two in number but ultimately six. They were chosen from patrician families when they were between six and ten years of age. They remained at their duties for thirty subsequent years and were expected to remain virgins all this time. If they failed to remain chaste, they could be entombed alive. Their duties consisted of maintaining the sacred fire, preparing sacrifices, and officiating at ceremonies.

While tending fires may not seem such an essential and sacred duty today, it is possible to think of many people of the past and present, of both East and West, who embody the sacred devotion to duty represented by Hestia. We see it in the Dalai Lama in the East and in many nuns and monks in both Asia and the Western world. Historical examples would include Saint Clare and Saint Francis. Saint Francis was a man who saw the world about him as sacred. He felt a kinship with all people and animals. In his "Canticle of the Sun," he speaks of Brother Sun, Sister Moon, Brothers Wind and Air, Brother Fire, Sister Death, and Sister Earth, our mother. Much of his attitude toward the world, people, and life is captured in his well known prayer:

> Lord, make me an instrument of thy peace.
> Where there is hatred let me sow love;
> Where there is injury, pardon;
> Where there is doubt, faith;
> Where there is despair, hope;
> Where there is darkness, light;
> Where there is sadness, joy.

O Divine Master,
Grant that I may not so much seek
To be consoled as to console;
To be understood as to understand;
To be loved as to love;
For it is in giving that we receive;
It is in pardoning that we are pardoned;
It is in dying that we born to eternal life.

As a small child, I heard stories in Sunday school about a heavenly realm above that left me in awe. By the time I was nine, however, I had gained a greater understanding of the universe, I dismissed the tales as fiction, and I aspired to become an astronomer. Yet throughout my life, unable to settle on any traditional religious path, I have been a spiritual seeker, searching for an understanding of life and the cosmos that lies beyond my grasp as a mere human. I am puzzled by the naivete of fundamentalists who take biblical myths literally, evidently failing to understand what they read. I believe rigid adherence to dogma is a barrier to true spirituality. (But one could argue that these are two alternative forms of piety.) I feel drawn to Eastern perspectives like Buddhism, which place less emphasis on mere belief than Western religions do.

I have long been fascinated by the lives of people like Saint Francis, who have actually lived their faith instead of imposing dogma on others. In my opinion, Saint Francis represents what was best in medieval Christianity, while the leaders of the Roman and Spanish inquisitions represent the worst. When I wrote my second novel, *Shaul of Tarsos*, based on the life of Saint Paul, I was not trying to make a religious statement, either for or against the ideas of the man. I felt a desire to get inside his head, to know what he experienced (especially the moment of his conversion), and to understand his view of life. I gained a deep appreciation of Paul in doing this, though I might not have liked him had I been around him in the first century. (Of course, the task of a novelist is like that of an actor. To treat Paul as the viewpoint character in my novel, I had to find a part of myself that matched that character, so I could think as he thought and feel as he felt.)

What place do religion, piety, and spirituality have in your life? Do you feel you have something in common with beings or people known for their piety? Can you think of such people you have known with whom you identify? To what extent do you regard piety as a part of your own nature? Is this a quality you feel a desire to cultivate?

9. Other Modes of Experience and Expression

In the preceding chapter, we considered several modes of experience, and we found that for each there were characteristic forms of expression. There are some additional modes of experience and expression that have received particular emphasis in mythic figures and in the mythic images that we project onto public figures.

EXPRESSIVENESS

Some mythic figures are known for the specific gift of fluent, eloquent, and persuasive speech. We have noted some of these in connection with their accompanying attributes. As a god of poetry in Scandinavian legends, Bragi was also known as a glib and witty speaker, perhaps a prototype for stand-up comedians. The Hindu god Ganesa, as a god of literature, and the goddess Sarasvati, as a goddess of knowledge and writing, are both associated with written expression.

In Celtic mythology, Ogma was a god of literature and eloquence. He was sometimes called Cermait, which means the "honey-mouthed." He is credited with inventing the Ogam alphabet, which originated in Ireland and spread to neighboring regions. It is composed of dots and lines, both vertical and slanted, and was initially intended for use in inscriptions on pillars.

The great communicator of Greek mythology was Hermes. As the messenger of the gods, he was a god who had a facility with words. He traveled to many places to impart the wishes of Zeus to mortals and to other gods and goddesses. When Zeus decided that Persephone should be allowed to return to her mother Demeter, it was Hermes who went to the underworld to inform Hades. In the modern world he could be a superb salesman.

Many writers have displayed a gift for expressing ideas and feelings in writing. William Shakespeare is an obvious example, but much could be said of the many other people who have demonstrated a talent for communicating effectively in poems, plays, novels, and political and social treatises. Many orators have been acclaimed for eloquent and persuasive speech. In classical Greece and Rome, this would include Demosthenes and Cicero. Many political leaders over the centuries have been known as articulate speakers. Abraham Lincoln could be eloquent using plain and simple language. Among presidents of the twentieth century, perhaps Franklin Roosevelt was most noted as an effective speaker. But some would argue that no American of that time was quite as eloquent as Winston Churchill in England. Ronald Reagan was often called the "great communicator" by his ardent supporters, perhaps because he used slogans and phrases (e.g., "evil empire") that appealed to them.

Some courtroom lawyers have been known for declamatory skills. Clarence Darrow, the eminent defense attorney, is a case in point. We have no recording of his courtroom performances, but Jerome Lawrence and Robert Edwin Lee have provided a fictionalized account of his role in the Scopes trial of 1925 in their play *Inherit the Wind*. Many women have proven to be effective speakers in the classroom, the courtroom, and on the political stage, though they have not commonly received the kind of public attention accorded the men I have noted, but that situation is certain to change in the years to come. Some women like Oprah Winfrey have shown great skill on TV as talk-show hosts or newscasters. In Shakespeare's *The Merchant of Venice*, of course, a lady by the name of Portia assumes the role of a persuasive lawyer.

In this section I have focused on expression in writing and speech, but we should recognize that many people are adept at non-verbal communication. Many actors and mimes have displayed considerable talent for conveying feelings and impressions through gestures, facial expressions, and body movement. Marcel Marceau and Robin Williams are outstanding examples.

I enjoy expressing myself in writing, and I have occasionally given effective speeches, but I have never considered myself a master of fluent oral communication. I doubt that I could be a very effective salesman. Perhaps if I had taken courses in speech and debate, I would have developed more facility at impromptu public speaking. Among the things I could do with the remaining years of my life, this possibility does not rank high on my list of unfulfilled potentials. But what about you?

Do you feel you have something in common with any beings or people noted for their ability to express themselves effectively in writing or speech? Can you think of such people with whom you identify? To what extent do you regard expressiveness, in writing or speaking, as a part of your own nature? Is this a quality you feel a desire to cultivate?

WILDNESS AND IMPULSIVITY

Many gods and goddesses have been associated with the natural realm. As a huntress, Artemis is associated with nature and with wild animals. Yet she is not seen as wild and impulsive herself. Within her domain, she maintained self-control.

The Greek deity who most clearly represents wildness and impulsivity is Dionysus. He is the subject of many legends and they vary from place to place. His origin as a mythic figure is disputed, but he may originally have been a god of wine. There are various stories of his birth and childhood. According to one story, he was the son of Zeus and Semele, a mortal woman. When Semele was pregnant, Zeus was tricked by Hera into revealing himself to Semele in his true divine form. When he did so, she was struck dead. Zeus then tore their unborn son from her womb and planted the fetus in his own thigh. Hermes later served as midwife for the birth of Dionysus.

To protect Dionysus from Hera's wrath, Zeus left him in the care of Semele's sister, and Dionysus was reared for a time as a girl. He was later left in the care of nymphs on Mount Nysa. As a child, he learned the making of wine and the pleasures of imbibing. Wherever he went, wine flowed, and he is associated with intoxication and orgiastic revelry.

Even as an adult, he often assumed the guise of a woman. He under-stood women, and as a rule, he treated them gently. He was worshiped by women, and he gave them freedom to experience their own sensuality. On the other hand, he could readily evoke terror and madness in people who sought to restrict his own freedom. He was at times accompanied by satyrs, the rustic god Pan, centaurs, and nymphs. He was, in all, a god who was in close touch with the feminine realm and associated with unbridled emotional states ranging from ecstasy to terror.

Most of the other creatures characterized by wildness and impulsive actions that we find in myths and legends have fewer human qualities than does Dionysus. Yet human figures of this type do appear in fiction. An enduring example is the well known creation of Edgar Rice Burroughs — Lord Greystoke, who was born into an upper-class family but chose to live in the jungle as Tarzan of the Apes.

In the realm of public entertainment we can find a number of people who manage to project an image of impulsivity and unconventionality. This would include Elvis Presley, the Rolling Stones, and the Beatles in the early stage of their work. Some of the subsequent stars of pop culture, like Michael Jackson, project an intersexual image that is reminiscent of Dionysus.

Many clowns and comics are adept at displaying a wild, unpredictable quality. One actor who has a remarkable talent for doing this is Robin Williams. We see this in such movies as *Mrs. Doubtfire* and *The Fisher King* and in his role as Mork in the TV series *Mork and Mindy*. We see it also when he appears for an interview on TV and proceeds to move about the stage, assuming a rapidly shifting set of roles. Pierce Brosnan, who was a member of the cast for *Mrs. Doubtfire*, once said that "when you're working with Robin Williams, the main thing you have to do is get out of the way."

Personally I have always led a fairly orderly life. As a child, I was too shy to engage in much wild or impulsive behavior. On rare occasions when I have done something very spontaneous, acting on the impulse of the moment, I have found it refreshing. Do you think I should do that sort of thing more often? Please bear in mind that if you encourage me to do so, I may just plan systematically to do something impulsive, and that will effectively foil the intention. And what about you and your life?

Are there people or beings noted in this section with whom you feel you have something in common? Are there people who act in wild or impulsive ways with whom you identify? To what extent do you regard impulsiveness as part of your own nature? Is this a quality you feel a desire to cultivate?

SENSUALITY

In Greek mythology, Dionysus was associated to some extent with sensuality and sexuality. But a god for whom this realm was more central to his role was Eros, often labeled the god of love. He represents love in both a sexual and a romantic sense. He carried a bow and arrows and was able to stir passion in people by shooting them with his arrows, leading them to experience both the joys and the anguish of love.

In the best known story of Eros, he succumbed to the passion that he usually aroused in others. The beautiful goddess Aphrodite was jealous of the princess Psyche, who had come to be known for her own beauty, and decided she had to dispose of the young woman. She commanded Psyche's father, on threat of dire catastrophes, to abandon his daughter on a high mountain where she would be the prey of a monster. Aphrodite then instructed her son Eros to dispose of Psyche by making her fall in love with the most vile and despicable creature possible. Eros, however, was so struck by the beauty of the maiden when he saw her that he fell in love with her himself.

In her sleep Psyche was whisked away to a palace, where Eros would visit her in the dark of night. He told her that he was the husband for whom she was destined, and he treated her so tenderly and lovingly that she looked forward joyfully to their nightly intimacies. Yet he asked her to promise that she would never attempt to see him in the light. When Psyche told her two older sisters of her wonderful new life, however, they were jealous and told her that her husband must be a hideous monster if he would not let her see his face. Given the doubts they aroused, Psyche lit an oil lamp one night as Eros lay sleeping. What she saw was not a monster but a remarkably handsome young god. Alas, a drop of oil lit on Eros's shoulder. He awakened, scolded her for breaking her promise, and suddenly vanished. The separation was painful for both lovers. Aphrodite subjected Psyche to a series of ordeals, which she managed to overcome.

Eros himself sought the help of Zeus, who made Psyche immortal, enabling Eros and his beloved to be reunited.

The Greek goddess most clearly associated with sensuality was Aphrodite, the goddess of love. In this role, she assumed forms that varied with place and time. In one form, she represented pure and ideal love. At another time, she was a goddess who favored and protected marriage and aided unmarried girls and widows in obtaining husbands. In another form, she was a goddess of lust and venal love and a patroness of prostitutes.

She manifested a beauty that inspired passion in the beholder. She took delight in arousing desire on the part of the other deities, and nearly all of them were thus launched on amorous pursuits with one another if not with her. Even Zeus succumbed to her charms, and this may have led to some of his encounters with mortal women. Aphrodite, of course, had her own share of intimate encounters with both gods and mortal men.

Aphrodite could be playful, mischievous, and vengeful at times. She could also be benevolent, as we see in the legend of Pygmalion. Pygmalion was a sculptor who lived on the island of Cyprus. He lived a solitary life, devoted to his art. Though he shunned the women of his region, he worshiped Aphrodite. In the course of his labors, he produced an ivory statue of a woman so beautiful that he fell in love with her. Alas, the cold statue did not respond to his embraces and kisses. But Aphrodite observed the scene and took pity on the poor man. As he took the statue in his arms, the ivory maiden began to return his kisses. Aphrodite had brought his work of art to life.

Cults devoted to Aphrodite appeared in many parts of the Greek world. At Paphos, the capital of Cyprus, there was a temple devoted to Aphrodite in the first century. It was there that young women preparing to be wed would come to sacrifice their virginity with strangers. Men would pay the temple for the honor of deflowering these young women. At other times, the temple priestesses served as sacred prostitutes.

Among the lesser divinities associated with sensuality were the nymphs and satyrs. The nymphs were young, charming, beautiful, and amorous divinities and were associated with various places in the mountains, valleys, and forests. The satyrs were more frightful in appearance and more lascivious and playful. They enjoyed music and dance and took delight in chasing the nymphs through the woods.

The sirens of Greek legends were sea nymphs known for their seductive charms. Originally they were described as having the heads and busts of women and the bodies of birds, and they were capable of flight. In later centuries, they assumed the form we recognize as that of the mermaid — the body of a women ending in a fish tale instead of legs. They dwelt on a tiny island surrounded by rocks, played either a lyre or double flute, and sang. Sailors who ventured near their island would be enchanted by their voices and, helpless to resist them, become shipwrecked on the rocks.

The Lorelei is a siren in a German legend of recent centuries. She lives on the river Rhine and lures fishermen to their destruction. She was the subject of a novel by Clemens Brentano and made famous in a poem by Heinrich Heine.

In the realm of public figures, it is possible to think of many people who manifest some sort of sexual charisma. That is, they have become objects of popular projections of erotic mythic images. Many movie stars of past and present come to mind. Such actresses as Hedy Lamarr, Elizabeth Taylor, and Sophia Loren all became known for exceptional beauty, and Brigitte Bardot, Jayne Mansfield, and Marilyn Monroe were regarded as having unusual sex appeal. On the male side, Rudolph Valentino was the unrivaled recipient of worship by women in the silent film era. Many subsequent actors have had their share of adoring fans — Clark Gable, Cary Grant, Rock Hudson, Paul Newman, Clint Eastwood, and Brad Pitt. If none of these actors or actresses suit your own tastes, you can add to the list. Perhaps you prefer Halle Berry, Meg Ryan, Marlin Brando, or Denzel Washington. We could also add the names of numerous pop music stars who have received their own share of adulation.

I confess that romantic love has been an important element in my own life. At times, in my youth, it was a source of obsessive preoccupation. I have also found that the nature of my love has gradually changed over the years, with a decreasing tendency to project an unreal image onto the object of my affection and a greater acceptance and valuing of the woman I have come to know. For me, as for most people, sex has also played an important role, but I shall spare you the details.

What about you and your life? As you contemplate the people and divinities we have touched on in this section, you will probably not imagine yourself playing the characteristic role of an Eros or Aphrodite — causing other people to fall in love with each other. Yet there might be times when you have sought to bring together a couple of people who strike you as an ideal match. The broader question here is the extent to which love and eroticism play a key role in your life. Is it important to you to be seen as attractive? Do you enjoy the company of physically attractive people? Is sexuality a central part of your life? Is it important to you to be in a love relationship, to be in love and to be loved? You can decide in what way love and sensuality play a part in your own life.

Bearing that in mind, are there beings and people we have noted with whom you feel you have something in common or who represent something important in your life? Are there lovers or attractive people with whom you identify? To what extent is love or sensuality a part of your own nature? Do you have a desire to cultivate such a quality?

THE CHILD

The child assumes several distinctive forms in mythology. One form is the divine child, the child seen from infancy as radiating a kind of magical, immortal quality. Then there is the precocious child, who displays unexpected talents from an early age. And above all, there is the child who displays the naturalness, innocence, spontaneity, imagination, and openness to fresh experience that we all had at one time and which too many of us have lost or covered up over the years.

Hermes as an infant is an example of a precocious and mischievous child. On the very day of his birth, he crept from his crib, climbed the mountains of Pieria, and stole some cattle that had been entrusted to the care of Apollo. From the divine herd, he selected fifty heifers and drove them to the banks of a river. In doing so, he made them walk backwards so the direction of their travel could not be discerned from their hoofmarks. After roasting two of the fattest cows, he returned to his crib. Apollo shrewdly determined who had committed the theft and brought Hermes to Zeus seeking divine justice. Zeus could only laugh and marvel at his tiny son's ingenuity, but he did order Hermes to return the stolen cattle to the herd.

Persephone represents the innocent child. Prior to being seized by Hades, she was gathering flowers in the meadow. Seeing a beautiful narcissus, she rushed over with delight, ever open to the wonders of nature. As she reached to pluck it, the earth opened up and she was then in the grasp of Hades. If you will recall the story, she was ultimately reunited with her mother Demeter, but she had to spend a third of every year with Hades — the time when the crops-to-be lie dormant in the ground. The rest of the year she was a growing girl, known both as Persephone and as Kore, the Greek word for girl.

As Kore, Persephone is a *puella aeterna*, ever young, spontaneous, and innocent. In some respects, her male counterpart is Dionysus, a *puer aeternus*, who forever displays the naturalness and spontaneity of a young boy.

Depictions of the child are far better represented in folk tales and fiction than in classical mythologies. In L. Frank Baum's *Wonderful Wizard of Oz*, in the popular novels of Lewis Carroll (*Alice's Adventures in Wonderland* and *Through the Looking-Glass*), and in the series of Harry Potter books by J. K. Rowling, we see children open to the wonders of the world and ready to explore magical realms beyond the mundane settings in which they live.

In the English tale of "Jack and the Beanstalk," we find a boy who has done something foolish — exchanging the cow for a handful of beans — but in his imagination, this ceases to be a cause for shame, for he is able to conjure up a magical world in which he overcomes a dreadful giant and riches ensue. In the German tale of "Hänsel and Gretel," two lost children who initially focus only on the immediate gratification of their needs undergo a bit of growth. They learn how to overcome the problems they face and return home.

Peter Pan, a character created by James M. Barrie, is a clear literary example of the *puer aeternus*. He enriches the lives of children by taking them on adventures into Never-Never Land. Of course, the children continue to grow up, but Peter remains ever a boy, moving in and out of a magical realm. Perhaps the pop star Michael Jackson is a living *puer aeternus*. He is fond of children and enjoys entertaining them, and he relishes opportunities to share experiences with young men who, like himself, have faced the problems and burdens of being thrust into the limelight at an early stage in life. He resembles Dionysus to the extent that he projects an image that is both childlike and intersexual.

Some writers have shown skill at entering the fantasy world of childhood and spreading it forth on the printed page. I think of such writers as James Barrie, A. A. Milne, Lewis Carroll, and Beatrix Potter. To produce such remarkable literary works, they must stay in touch with the child within themselves, but they may also benefit from being around children with whom they can share their fantasies.

I have a lifelong habit of scanning back over my life and trying to make some sense of it all. In my memories, I can return to any point in childhood. I can return to the child I was at age one, two, or three and relive the wonder of looking at the sky or seeing tiny creatures crawling through the grass. I remember a day when my older brother and I were walking up the street behind our parents. Passing a weed-strewn vacant lot, we spotted a large chunk of plaster on the ground, white on the rough edge with a bit of color on the smoother side. My brother said to me, "That's a piece of the sky." I felt a deep sense of awe, for indeed, a piece of that vast blue ceiling far above us had obviously fallen to the ground.

I was destined to gain a much different understanding of the sky and celestial bodies that lie out there in the succeeding years, but I have often returned to that early moment as a reminder of the wonder and awe that are the natural response to the mysteries we two-legged mammals face in trying to grasp the meaning of our existence. We can accomplish much through rationality and empirical science, but the greatest mysteries will always elude our grasp. At best, we can arrive at a sense of connection with them by creating myths and art forms — in painting, sculpture, music, poetry, drama, and dance. We gain a momentary satisfaction, but we must continue creating, for we never reach an end-point in our quest for meaning. Once, as I returned to that early moment in my life, I wrote the poem "Older Brother" and at a later time used the words in an art song:

When you were three and I was two
And we walked way down to the end of the block
To explore the mysteries of a vacant lot,
What was that chunk of stuff upon the ground,
all pink and blue?
You knew of course and told me,
And sensing the wisdom of your years
I knew you knew,
For I had seen those colors too
way up above.
That was long ago and years go by;
Yet I shall not forget —

When you were three and I was two
We found a piece of sky.

I believe that there always remains within each of us a child that we cover up as we learn fresh ways of acting and thinking while moving through the later stages of childhood, adolescence, and the adult years. People vary in the extent to which they remain in touch with the inner child and in the extent to which they believe it is important to do so. For some, it seems important to think and behave like a responsible adult every moment of the day. Others appear unable to outgrow habits they acquired early in life, and they continue playing roles that become increasingly maladaptive as they pass through adolescence into adulthood. Either extreme suggests a lack of desirable flexibility. Yet we need not all make the same choices. What are yours?

Are there people, beings, or fictional characters we have noted with whom you think you have something in common? Are they people who manifest some facet of their inner child with whom you identify? To what extent do you consider your child side or some expression of it as an important part of your own nature? Do you feel a desire to cultivate such a quality?

The Trickster and the Psychopomp

The psychopomp, the trickster, the clown, and the shaman are figures that turn up in a range of cultures and mythologies. However varied their form as people or as mythic beings, they share some common characteristics. Each one shows an inclination to transcend the boundaries that most of us do not cross — the boundaries between conscious and unconscious, between the child and the adult, between ordinary reality and fantasy, between life and death.

A psychopomp is one who conducts souls to the place of the dead, and this was a primary role of Hermes in Greek mythology. He was also the being who could most readily move from one realm to another, going from Olympus to the sea, the underworld, or the realm of living mortals. He could control the transition between sleep and waking. He induced sleep in the Greeks so that Priam could retrieve the body of his son Hector from the battlefield outside the walls of Troy.

Anubis (or Anpu) plays a corresponding role in Egyptian mythology. It is he who guides souls about the underworld and to the realm of Osiris, lord of the underworld. It is said that he accompanied Osiris and aided his efforts to conquer the world. He also helped Isis prepare Osiris for burial and has therefore been associated with funeral rites and was known as "lord of the mummy wrappings."

The trickster is a figure that appears in the folklore of African tribes and various indigenous people of North and South America. Such a figure is usually male, but sometimes female. It may be part animal, part human, part divine, appearing now as one and now as another. The figure could cross over the boundary separating the corporeal from the spiritual. The trickster is often a mischievous or amoral troublemaker, but he can also do something beneficial for others. We see this in the case of Raven, a trickster of the Pacific Northwest. Raven made himself very tiny and entered the body of the daughter of a chief. He later emerged disguised as a newborn infant and then stole the box in which the chief had hidden the sun. He thus managed to bring light to the world. We should all be grateful.

Odysseus is a Greek hero who displays some of the traits of the trickster. In some tales he is depicted as wily, devious, and deceitful. In Homer's *Odyssey* this trait appears as a resourcefulness that enables him to escape from difficult situations. A key event occurs early in the journey he undertakes to return from the Trojan war to his home in Ithaca. His ship stops at the island of the one-eyed giants known as cyclopes, and there he encounters the cyclops Polyphemus. In the course of their interaction, he identifies himself as Nobody. Polyphemus imprisons Odysseus and his men in a cave where he keeps his sheep. Odysseus, however, manages to get the cyclops drunk and then put out his one eye with a sharpened, burning stake. He and his men escape from the cave by tying themselves under the bellies of rams and flee to their ship. Upon seeing the blinded Polyphemus, the other cyclopes ask, "Who has done this to you?" They would have given him assistance, but he answers that "Nobody has done this." Left alone in his rage, Polyphemus hurls a large rock at the ship as it sails away. Then he prays for his father, Poseidon, to prevent Odysseus from returning home.

For ten more years, Odysseus makes many detours and faces more obstacles in the way of his homeward journey. At last, Athena asks Zeus to intervene and allow him to return to Ithaca. On the surface, Odysseus appears in this tale to be a clever trickster, but another reading on a symbolic level is possible. One could argue that in his encounter with the cyclops, he does indeed lose his conscious identity and ceases to be the man he was. He plunges into another world, a world known to psychotics, where he must undergo a long struggle before he can return to ordinary reality. Finally rationality emerges and intervenes. He lands on the island of Ithaca, and initially only his faithful dog recognizes him. The animal, having awaited his return for many years, leaps for joy and falls dead at his feet. Odysseus must still work his way through the human realm and fully re-establish his true identity in order to be reunited with his wife.

The kind of psychotherapist who can work most successfully with schizophrenics is one who can enter into the world of the psychotic patient and understand what that patient perceives and how he or she feels about it. In a sense, such a psychotherapist is the modern equivalent of the shaman. The shaman (sometimes called a medicine man or witch doctor) is a religious practitioner found in tribal groups or small communities of North and South America, as well as in Siberia. The shaman is believed in those communities to be able to diagnose and cure illness by means of a special relationship with the spirit world that surrounds us but remains unseen by most of us. On occasion, the shaman may choose to cause illness. Shamans commonly accomplish their work by entering a trance, and they may do so by means of autohypnosis, hallucinogenic substances, fasting, or self-inflicted pain. In this manner, the shaman moves temporarily to a different state of consciousness and may act or speak in ways we might consider psychotic. Yet from the standpoint of his own culture, he is able to gain insights and accomplish deeds not possible for others.

The shaman crosses a boundary between states of consciousness. The clown crosses the boundary between adult and child. Upon donning his — or her — costume, the clown entertains us and sometimes entices us to follow suit by displaying the spontaneity and antics of a child whose behavior has not yet been curbed by parents and teachers. A great actor or actress must also be a crosser of boundaries. To do a successful job of playing many different roles — say, a schizophrenic, a mentally deficient person, a diabolical fiend, a carefree ne'er-do-well — one must enter into the world of that character and sense his or her inner thoughts and feelings.

As a psychologist I have long had a fascination with different states of consciousness and with the different contents associated with them. I have gone through periods in which I have paid special attention to my dreams and made an effort to become aware while dreaming that I was dreaming. I have noted moments in dreams and in relaxed waking consciousness when I seemed to reach out through time or space and be in touch with other people or subsequent events via telepathy or precognition. I have enjoyed playing different roles with children and with adults with whom I feel close. While I have taken a few acting classes, however, I have never devoted much time to taking part in dramas. Perhaps this is a talent I can cultivate in another lifetime.

What about your life? You may not think of yourself as a psychopomp, trickster, or shaman. Yet we all move from one state to another as we pass between sleep and wakefulness. Perhaps you have entered other states as a result of consuming alcohol or some other drug. Maybe you have done so by means of meditation or hypnosis (self-induced or induced by someone else). We often enter different states without deliberately seeking to do so. We may undergo a bit of a shift in roles when we take part in a play or in a more ordinary social setting. We may experience a shift within when we empathize with another person whose life is different from our own. We may also discard some of our usual social inhibitions when we play with children or animals, becoming for the moment a child, beast, or court jester. Do you find yourself crossing any boundaries — between child and adult, normal and "crazy," civilized and wild?

Do you feel you have something in common with any of the beings or types of people we discussed above? Are there such people, crossers of boundaries, with whom you identify? To what extent do you consider this quality a part of your nature? Is it a quality you feel a desire to cultivate?

10. Developmental Paths

In the preceding chapters, we have considered a wide range of mythic figures. I believe they embody archetypal images of the major modes of acting, experiencing, and being that exist as potentials for all of us. As I have classified them, they include the following basic themes:

(1)   nurturance and compassion
(2)   ascendance
(3)   aggression
(4)   autonomy
(5)   ordered, rational thought
(6)   intuitive wisdom
(7)   aesthetic and imaginal pursuits
(8)   piety
(9)   expressiveness
(10)   wildness and impulsivity
(11)   sensuality
(12)   the child (openness, innocence, spontaneity)
(13)   the trickster and the psychopomp (transcendence of boundaries)

We often think of some of these modes as either masculine or feminine in quality. As I have indicated, however, it is possible for each of them to find both male and female representatives, whether we are looking at mythic figures or historical examples.

If all of these modes are universal potentials, does this mean that each of us should cultivate all of them? I would hesitate to make that claim, but I believe it is valuable to recognize all of them as possibilities. It is all too easy to focus exclusively on a single mode, only to find that it does not meet all our needs. Over-concentration on that one mode may, in fact, prevent us from meeting many needs. Our lives then become very dissatisfying, and we may feel caught in a trap that affords no escape. The escape may lie in recognizing the avenues that become available when we cultivate an alternative mode.

Is there a specific course that either men or women should follow as they seek to realize more of their potentials while progressing through their adult years? Many writers appear to believe so, and it is useful to consider their proposals. In this chapter, we shall first examine the ideas of writers who believe that contemporary society limits the ability of men to realize their full masculine nature. These writers try accordingly to highlight the potentials that men need to realize. Then we shall examine the work of writers who focus on qualities that women need to realize. Finally I shall note some psychoanalytic and Jungian views regarding stages of psychological development. I see value in the contributions of all the writers I discuss here, but I believe that a consideration of all the basic themes I have noted above may provide the foundation for a more comprehensive picture of psychological development in the adult years.

Full Realization of One's Masculine Nature

In recent decades, many people have stressed the importance and value for men to get in touch with the feminine side of their nature. One writer who has sounded a cautionary note, however, is the poet Robert Bly (1990). While he does not believe that men should deny their feminine qualities, he believes that there is a greater need in our society for men to develop their masculine potentials. He claims he has encountered too many men since the 1960's who have developed feminine qualities of sensitivity and nurturance but who lack resolve and decisiveness because the masculine side has been neglected and devalued.

Bly contends that a basic problem in our society began to develop late in the nineteenth century as men began to move from farms to factories. Increasingly since that time, the father in the family has gone off to work in the morning and remained out of sight of the son until returning in the evening. In an earlier era, the son would have participated more and more in the work of the father in the course of growing up. Now the father remains a more distant figure. Without the kind of bonding and identification that naturally occurred at an earlier time, the son grows up more passive and naive.

Other societies employ various practices for initiating boys into the world of men. They may remove the boy from the dwelling of the mother at a certain age and place him in a dwelling for men, so that he is interacting primarily with them from that point on. The boy may be subjected to a period of initiation in which he is wounded, placed in a pit, enslaved, or made to lie in ashes. Some of these practices appear in myths and legends that come to us from earlier times.

Bly believes that a boy growing up in a typical American or Western family is ill prepared for manhood. Furthermore, the images of manhood provided by popular culture and presented in our highschool years are inadequate — images of righteousness, toughness, and achievement in various forms. The potentials that boys and men need to embrace are revealed in archetypal images that recur in myths, legends, and fairytales. Bly's work has inspired a men's movement in which men seeking to realize their masculine potentials gather in groups, undergo rituals, discuss their lives, and offer a bit of mentoring or guidance to one another.

According to Bly, men traditionally have gone through a number of stages in development that may be incorporated into initiation rites. Initially there is a period of bonding with the mother, followed by separation from the mother. Then there is a period of bonding with the father, followed by separation from the father. Bly believes that for men in our society all the stages that follow the initial mother bonding tend to be derailed or not realized at all, so that the growth that should follow these early stages remains unrealized.

In the course of his book *Iron John*, Robert Bly illustrates the points he makes regarding the needs of contemporary men by telling an ancient story one step at a time, the tale of Iron John. Iron John appears to the tale's hero through most of the story as the Wild Man. He is a figure at odds with the surrounding society and separate from it, but he serves as a guide and mentor to the young hero, who grows thereby to full manhood. Bly argues that each of us men needs to get in touch with the wild man within, and we also need to bring the inner warrior to life. To put it another way, every man needs to transcend whatever inadequate or misleading directives our family and the surrounding society have offered us and find the natural directives within our own basic nature.

In a further elaboration of his position, Bly speaks of seven beings within a man's psyche with which he needs to get in touch. These are the King, the Warrior, the Lover, the Wild Man, the Trickster, the Mythologist or Cook, and the Grief Man. The "Grief Man" strikes me as a different sort of image from the others, since grief does not seem the most salient feature of any mythic figure. To be sure, some mythic figures go through periods of grieving, and we have noted this with respect to Orpheus and Demeter. It is certainly important for each of us to be able to experience grief, rather than forever suppress the experience. I agree with Bly that men must allow themselves to grieve. The other six beings correspond to modes I have discussed in terms of ascendance, aggression, sensuality, wildness, the trickster, and the psychopomp or shaman.

I can certainly agree with Bly that there is value in being in touch with a range of potentials, but I note that his list does not include any potentials that might be labeled feminine. Rather than embracing androgyny as a desirable goal, Bly cautions against a merging of the man and the woman in a marital relationship. The woman should not become an extension of her husband's personality, nor should the man become an extension of his wife's. There is value in the tension created by the difference between them. As opposites, they complement each other.

I believe that the truth of such a generalization depends on what each partner brings to the relationship. I agree that either partner who becomes submerged in the shadow of the other partner loses something in the process, but the marriage between a man stuck in an overly aggressive role and a woman stuck in a nurturing role can be even more detrimental for the growth and welfare of both partners. A marriage can be rich if each partner has flexible access to a range of potentials. If so, it may not matter whether as individual composites they look like "opposites" or as similar androgynous mixtures of traits. If each partner possesses a rich personality, they can always bring contrasting views to bear on any issue within their relationship.

Robert Moore and Douglas Gillette (1990) present a position that is similar to that of Robert Bly. They believe that most men in our society manifest a "boy psychology," because their masculine potentials remain at an immature level. This immaturity is expressed in many ways — e.g., in abusive and violent actions, weakness, passivity, inability to act effectively or creatively, and oppression of women or anything feminine. Like Bly, they believe that the primary problem of men today is not a lack of connection to the feminine but rather a lack of deep connection to the masculine. This lack in turn complicates their relation to their own feminine sides, as well their relation to women.

Moore and Gillette see a value in being connected to the inner child. In the early development of the boy, the child archetype appears in several forms, and our connection provides a basis for creativity, playfulness, wonder, and curiosity. There is a value in experiencing a connection to the inner child throughout life, but it is important for a boy not to be identified with any form of the child archetype. As a result of such an identification, according to Moore and Gillette, he may play the role of a know-it-all, a coward, a bully, a high chair tyrant, a dreamer, or a mama's boy.

These authors also speak of four mature forms of the masculine to which every man needs to relate. These include the King, the Warrior, the Magician, and the Lover. The archetypal foundation for all four of these is part of our genetic programming. Unfortunately they may all yield undesirable effects if they are manifested in immature forms. As with the child archetype, it is important to experience the connection to each archetype and allow it to manifest in a balanced relationship to the other archetypes. The immature or shadow form of the archetype arises when we identify with a given archetype and become possessed by it.

Moore and Gillette speak of the King as the central archetype and a source of father energy. It provides the qualities of order, rational patterning, and integration and integrity in the masculine psyche. Properly realized it ensures a perfect balance for the rest of the archetypes. In shadow forms it is manifested as the Tyrant or the Weakling. Over-identification with the King archetype has long been a feature of Western and Middle Eastern civilization, and we see it expressed in patriarchal rule and the subjugation of women.

The Warrior archetype can serve as a source of energy and motivation. It is ideally expressed in a transpersonal commitment to a god, a cause, a people, a nation, or a heroic task. Directed into more personal, self-serving channels, it is manifested as the Sadist or the Masochist. Violence for its own sake or for sheer personal satisfaction is an immature expression of this archetype and responsible for a widespread negative view regarding the Warrior image.

The Magician archetype is a source of thoughtfulness and reflection. It provides the energy of introversion and access to the inner world most available to introverts. Moore and Gillette say that the specialty of the Magician is "knowing something that others do not know," and it is the province of men who have been known as shamans or holy men. A man who realizes this archetype in a mature way would use specialized knowledge in a manner that would benefit others. Realizing it in an immature form, he might use his expertise in a non-constructive or destructive manner, acting as a detached manipulator.

The Lover archetype provides the basic energy that enables us to feel fully alive and to experience passion and vivid sensations. Moore and Gillette consider this an archetype that guides psychics and inspired people in various artistic fields. A man possessed by this archetype may feel overwhelmed by sensations or become addicted to drugs, sex, or various kinds of objects or sensations.

Many of us would like to believe that patriarchal rule is on the decline and that our society is moving toward greater equality of men and women. We would also like to believe that violence is also gradually declining, even if American voters occasionally elect leaders who choose to launch unnecessary wars. If these changes are indeed progressing, Moore and Gillette would probably ascribe them to a greater balanced realization by men of the four masculine archetypes. I would suggest, as an alternative perspective, that we might regard this kind of progress in terms of a greater realization by both men and women of alternative modes of functioning. Such a realization enables men to rely less on an assertion of power and women to experience more of their own power.

In addition to the male authors we have considered above, we may note books dealing with Greek gods written by two women authors, Jean Shinoda Bolen (1989) and Christine Downing (1993). Both of these authors take the position that the gods and goddesses represent qualities, or ways of being, that enter into the experience of both men and women. While recognizing that the gods are manifested more strongly in men, they are concerned with how those gods are experienced by women as well as men.

Bolen devotes a chapter each to the following gods: Zeus, Poseidon, Hades, Apollo, Hermes, Ares, Hephaestus, and Dionysus. While women may not experience these gods are strongly as men do, women experience them in various ways. They may find that the gods correspond to parts of themselves — e.g., when they display the cold and objective thinking of Apollo, the introverted creativity of Hephaestus, or the wild sensuality of Dionysus. On the other hand, understanding the gods may enable a woman to understand men better. She may perceive a quality of one of the gods acted out by a particular man, or she may recognize that her reactions to various men are governed by the mythic qualities she sees in them (leading her to be drawn to some men and repelled by others).

Downing also devotes eight chapters to eight different gods: Hades, Hermes, Dionysus, Apollo, Hephaistos (Hephaestus), Ares, Poseidon, and Zeus. Each of these gods may represent a quality that a woman perceives in someone else and that she relates to, but it may also represent a quality that she identifies with and experiences in herself. It is common to think of the gods as representing masculine qualities and the goddesses as representing feminine qualities, but Downing argues against this simple equation. She regards the bipolar view of masculine vs. feminine as distorted and arbitrary and believes it is not much better to regard the gods as representing different forms of the masculine. Of course, I have made essentially the same point by noting that each of the basic human qualities expressed in mythology is represented by both male and female mythic figures. And I would agree that the traits we tend to label masculine and feminine are subject to cultural variation and that our categorization of them as masculine or feminine is rather arbitrary. Nevertheless, biological differences between men and women also play a role, most clearly in the case of physical aggression.

REALIZATION OF ONE'S TRUE FEMININE NATURE

Just as Robert Bly says that men must get in touch with the Wild Man, Clarissa Pinkola Estés (1992) contends that women need to get in touch with the Wild Woman within themselves. She regards this as the most natural, instinctual part of women, and she also speaks of it in term of the wolf-woman Self. As the title of her book (*Women Who Run with the Wolves*) implies, healthy women and healthy wolves have much in common. Both are by nature keen-sensing, playful, capable of deep devotion, deeply intuitive, stalwart, brave, intensely concerned with their young and family, and unfairly maligned.

Estés draws tales from a variety of cultures to illustrate her points, but she notes that the Wild Woman is inadequately represented in fairytales and myths collected in the Western world in recent centuries. What might have been a wise woman guide in an earlier tale may have been transformed into a wicked witch in later versions. Of course, the very word *witch* (along with its equivalents in Old English and related Germanic tongues) acquired a negative connotation under the influence of Christianity as it spread from the Mediterranean northward through Europe. From time to time, women failing to bow to authority and suspected of thinking for themselves have been labeled witches and either burned at the stake or hanged.

As the term Wild Woman suggests, many of the images provided to women in our society — images of what is deemed desirable or ideal in a women — are images that serve to mask the most basic layer of a woman's nature. Yet we can find depictions of the Wild Woman in many ancient legends and myths. Of the mythic figures we have considered in this book, Artemis is the best case in point. Estés contends that women must allow this part of their nature to express itself and must be free to move, create, and be angry. She also stresses the importance of cultivating intuition as a way of being in touch with one's instinctual roots. Lacking this, a woman may simply let herself be guided by the rules, guideposts, and things provided by the people and community around her. Relying on intuition, she can find an inner source of guidance and make choices more appropriate to her actual needs. A woman must assert her right to be aware and to be natural and creative.

Madonna Kolbenschlag (1979) offers a related critique of the expectations imposed on women in contemporary Western society. She says that boys are encouraged to be active and independent, while girls are encouraged to be desirable, friendly, subordinate to boys and men, and more concerned with the needs of others. She believes that traditional myths and tales generally depict male heroes as progressing toward self-realization, but female heroes do not show a comparable development.

The tales to which she devotes most of her attention are Sleeping Beauty, Snow White, Cinderella, Goldilocks, Beauty and the Beast, and the Frog Prince. In the first three of these tales, we see a maiden endowed with great natural beauty who is never free to act on her own. She must simply wait for the prince of her dreams to discover her and marry her, enabling her to live happily ever after. Of course, these are all tales that have gained great popularity in recent centuries.

To the extent that such tales are representative of tales and myths presented to girls in Western society, Kolbenschlag is surely right in contending that they diminish women and serve to reinforce the limited role choices that our society has offered to them, and women need to transcend these myths to reclaim their full humanity. As a qualifying comment, I would merely note that we have seen a greater variety of roles displayed by the female mythic figures of the past and that it is possible to find many popular stories of recent vintage (e.g., Anne of Green Gables) that depict girls in a more active, adventurous form. In accord with Madonna Kolbenschlag, I welcome the ongoing progress toward a greater range of role choices available for women.

Both Jean Shinoda Bolen (1984) and Christine Downing (1981) have written books about Greek goddesses and heroines, endeavoring to show women the possibilities available to them as represented by archetypal images. Downing devotes the chapters of her book to the Great Mother (Magna Mater), Persephone, Ariadne, Artemis, Aphrodite, and the Child. In the preceding chapters I have touched on all these mythic figures except Ariadne, whom Downing regards as a psychopomp comparable to some extent to Hermes. The best known tale of Ariadne concerns her relationship with Theseus, a hero who has undertaken the task of slaying the Minotaur. This beast dwells at the center of a labyrinth, and Ariadne aids him by guiding him both into the labyrinth and out again once he has accomplished his deed.

Bolen devotes her chapters to Artemis, Athena, Hestia, Hera, Demeter, Persephone, and Aphrodite. She speaks of the first three as virgin goddesses who represent different forms of autonomy. Each foregoes a close involvement with a god or man so that she can maintain her independent course. The next three — Hera, Demeter, and Persephone — are more vulnerable because they are relationship-oriented (as wife, mother, and daughter). She views Aphrodite as an alchemical goddess, a goddess with the power of transformation, who can inspire creativity and love in all forms.

Bolen notes that women differ in their needs and that some may want to call on the energy of relationship-oriented goddesses, while other may be more inspired by images that embody independence or spirituality. Furthermore, a woman's needs may be represented by different goddesses at different stages of her life. Jean Bolen is a psychiatrist and Jungian analyst, but she points out that both Freudians and Jungians often manifest a simplistic view of the masculine and feminine. They often assume that a woman pursuing an aggressive independent course of action is acting out her masculine side (her animus), while in fact she may be expressing the energy of a feminine goddess like Artemis or Athena.

LIFE STAGES

Life stages have been postulated in various forms in myths, in the works of great writers, and in popular conventions. Shakespeare speaks of seven ages of man in "As You Like It." The legal notion that one becomes an adult at 21 rests on a quaint Western convention that human development proceeds in the three seven-year periods of early childhood, late childhood, and adolescence.

In the realm of psychological theory, the developmental scheme that has received the greatest attention over the past century is that of Sigmund Freud, though few psychologists today would claim much allegiance to the Freudian view. Freud postulated three early psychosexual stages — oral, anal, and phallic — followed by a latency period extending through the later years of childhood. Finally there is a genital stage. Ideally one reaches this in adolescence, but it is conceptualized in terms of full readiness for a mature sexual relationship with a partner of the opposite sex. Full development of genital sexuality, thus understood, may require several years of development and reach full blossom in the adult years.

Freud spoke of stages extending only through childhood and adolescence, but another psychoanalyst, Erik H. Erikson (1963), has modified the scheme and extended it to encompass eight life stages. He labels them as follows:

(1)   basic trust vs. basic mistrust
(2)   autonomy vs. shame, doubt
(3)   initiative vs. guilt
(4)   industry vs. inferiority
(5)   identity vs. role confusion
(6)   intimacy vs. isolation
(7)   generativity vs. stagnation
(8)   ego integrity vs. despair

Each stage is characterized in terms of a polarity. At any given stage, the individual is likely to experience both poles (e.g., both trust and mistrust), but the ideal outcome of any stage is a predominance of the positive pole (basic trust, autonomy, initiative, industry, etc.). To some degree the issues focal in a given stage will still be experienced in any subsequent stage, but they may be overshadowed by the issues more central to that later stage. If the individual passes through a given stage with a predominance of the negative pole (more mistrust than trust, more shame than autonomy, more guilt than initiative, etc.), that negative outcome is likely to create problems in later stages.

The first four of these stages correspond age-wise to the stages Freud called oral, anal, phallic, and latency. In our society a child would pass through the first three in the pre-school years and enter school in the fourth stage. Thus, the fourth stage is one in which the child is expected to learn the three R's and acquire the basic tools he or she must use to play the roles demanded by the surrounding society.

Stages 5 and 6 correspond roughly to Freud's concept of a genital stage. Erikson sees the basic issue in stage 5 as a developing sense of ego identity. This would include a sense of sexual identity — not the mere sense of early childhood that one is male or female, but an expanded sense of oneself as a sexual being in relation to male and female peers. It may also include an evolving occupational identity. Stage 6 is a stage of young adulthood, a stage in which ideally one can function as a responsible adult. In particular, this involves a capacity to commit oneself to concrete affiliations and partnerships — in love, marriage, friendship, and work.

The issue that is more central in stage 7 is a concern with guiding the next generation. For most people, this concern would focus on their own children and grandchildren, but it can extend beyond this. Many people with no children of their own manifest a high level of generativity. Some of them become teachers and have a lasting positive impact on hundreds of the children and teenagers who pass through their classes over the years. Others may serve as priests, nuns, counselors, or social workers. Or they may devote their lives to the care of animals or various groups of disadvantaged people.

Erikson thinks of ego integrity as a sense that one's life is a meaningful, orderly whole and that it had to unfold as it did. He suggests also that a possessor of integrity will regard his or her own life as but a coincidental segment of history in a given culture and that, experiencing oneself as a meaningful part of that culture in a given phase of history, one can more easily accept death. Erikson's rationale here is rather general and vague, but his viewpoint is decidedly optimistic.

In describing this scheme, Erikson does not deal with gender differences or with variables we might readily regard as masculine or feminine. He sees the central issue of early adulthood as the capacity for establishing mutuality in relationships. This is particularly important if one is going to enter into a lasting love or marital relationship. In the stage that follows, the key issue is nurturance, providing for those less able to care for themselves or to respond in kind. Nurturance is often regarded as a feminine quality, but in many forms it is displayed as often by men as by women.

Erikson may be considered a neo-Freudian. Jungian theory offers a somewhat different way of looking at development in the adult years, though Carl Jung never set forth a scheme of life stages. From Jung's standpoint, the major achievement of early adulthood is full development of the ego. As Jung defines this term, the ego is the center of consciousness. It is that part of the total psyche concerned with one's sense of conscious identity and with the attitude and functions one uses primarily in dealing with the world and life.

According to Jung, there is one basic attitude dimension — introversion vs. extraversion — and each of us tends toward one pole or the other. The thoughts and actions of the extravert are governed more directly by things, people, and events in the environment. The thoughts and actions of the introvert are governed more directly by subjective factors, and the response of the introvert in any situation may be more inward and less outwardly evident than that of the extravert.

There are two function dimensions: sensation vs. intuition and thinking vs. feeling. Sensation and intuition are considered "irrational" functions, and they concerned with the apprehension of events. The sensation function focuses on the immediate contents of experience. In any given situation, the sensation extravert focuses on the things and ongoing events. The sensation introvert attends more closely to the sensory effect itself and is more aware of colors, forms, textures, etc. Many artists are sensation introverts. Intuition is concerned more with the possibilities and meanings suggested or implied by the given situation. The intuitive extravert would be interested in concrete possibilities, while the intuitive introvert might be more interested in visions, fantasies, extrasensory readings, or the inner world of the person observed.

Thinking and feeling are both concerned with judgements. The thinking function is more rational in the sense that it categorizes the things observed and focuses on the relationships among them. Feeling is more concerned with value. Things are regarded as good, bad, beautiful, ugly, desirable, undesirable, etc. Of the four functions, the one on which the individual relies most heavily is called the superior function, while the function favored in the other pair is called the auxiliary function. The two remaining functions, those less utilized by the individual, will still operate but under less conscious control. For example, the individual who relies heavily on thinking may make value judgements but believe that he or she is simply drawing conclusions based on cold, hard reality.

According to Jung, the ego of the man is masculine, and the ego of the woman is feminine. Unfortunately, Jung never provided a detailed description of masculinity and femininity as he understood these concepts. He spoke of logos as the masculine principle and of eros as the feminine principle. *Logos* for Jung implies something in the way of objective interest and discrimination, while he speaks of *eros* as a principle concerned with relationship.

He contends that everyone has both a masculine side and a feminine side. While the man has a masculine ego, he also has a contrasexual side — a less conscious, feminine portion of his psyche called the anima. The woman has a feminine ego, and her less conscious, masculine (contrasexual) side is embodied in her animus. We are more likely to experience the anima or animus in projected forms than to recognize it as a part of ourselves. The man may experience anima images in the form of female figures in his dreams, and he projects such images onto women who seem to have a strong emotional impact on him that he cannot explain — e.g., a woman with whom he falls in love. The woman likewise experiences her animus in projected form in men who appear in her dreams and various men she encounters in waking life.

According to Jung, most of the psyche remains unconscious, and he speaks generally of the content that lies outside consciousness as the shadow. In both dreams and waking life, we may project images that contain various parts of ourselves that we have not consciously owned — e.g, the less favored functions and various aggressive and sexual impulses. While the ego is the center of consciousness, Jung speaks of the self as the center of the whole psyche. While the normal course of development entails full development of the ego in early adulthood, the ideal movement in the subsequent years is toward a realization of the self, and Jung speaks of this progression in terms of individuation.

Individuation does not imply full consciousness of all the contents of the psyche, though we inevitably become more conscious of much of those contents. It implies rather a greater recognition that much lies outside out consciousness and that we must allow it to be expressed. Among other things, this would mean that we pay more respect to the ideas, feelings, images, and dreams that pop into our heads and realize they have something to tell us. A related implication of this view of development is that we emerge into adulthood as lopsided beings, as either masculine or feminine, and relying heavily on one attitude and two particular functions. In the course of individuation we may come to make more effective use of the other attitude and functions. Furthermore, we will tend to integrate more of our contrasexual side into consciousness. In other words, we will become less lopsidedly masculine or feminine and more androgynous.

However we choose to define masculinity and femininity, it makes sense to think of androgyny as representing a wholeness toward which we might all reasonably strive. Furthermore, the lopsided development of the ego in early adulthood accords with the view of writers like Robert Bly, who shudder at the thought of young men cultivating their feminine qualities at the expense of their masculinity. Yet most of what we call masculine and feminine is subject to cultural variation. In some societies we see little behavioral difference between men and woman, while other societies demand a sharp division between the social roles and habits of men and those of women. There is a question I feel compelled to ask: in the absence of legally enforced gender roles, is there any reason why lopsided development in an early stage of adulthood is essential? Is it not possible to cultivate both masculine and feminine qualities jointly throughout life — in short to follow an androgynous course throughout development and attain the ultimate wholeness envisioned by Jung? I believe so, but I do not believe we all have to follow the same developmental path.

Jung's concepts of anima and animus remain a bit fuzzy in the absence of a clear articulation of their ingredients. If we think of each as representing a sort of "soul image," the anima of the man embodies, among other things, his contrasexual ideal, the woman he wants as a counterpart. The animus would embody the the contrasexual ideal of the woman, the man she wants as a counterpart. If so, homosexuals complicate the picture. The gay man seeks a male counterpart, but that may be a man whose qualities complement his own. The lesbian too may seek a woman whose qualities in some way complement her own. It takes but a brief bit of reflection regarding the homosexual and heterosexual couples we know to realize that what any given individual seeks in a partner may be easy to define with respect to sexual anatomy but not with respect to qualities readily identifiable as masculine or feminine.

11. Development Goals

In the preceding chapters, we considered various ways to progress in personal development by realizing potentials we have neglected or failed to recognize. We examined a number of modes of being embodied in mythic images that we encounter in ancient tales and often project onto public and historical figures. We also noted the views of writers who stress the need for men to realize their full masculinity and for women to realize their true feminine nature.

In this chapter, we shall take another look at ways in which we can overcome some of the obstacles in our path by cultivating potential modes of being that can balance ones we have over-emphasized. Writers who advocate the cultivation of androgyny have offered valuable ideas regarding a desirable balance. I believe, however, that casting the issue only in terms of "masculine" and "feminine" modes imposes an unnecessary restraint on our thinking, and I propose a broader consideration of flexibility and balance.

There also looms the over-arching question about whether there is one common goal to which we should all aspire in our personal development. We shall examine some of the possibilities. They include several basic modes of fulfillment embodied in ideals found in Eastern and Western traditions.

Androgyny as a Developmental Goal?

Jung did not explicitly speak of androgyny, but other Jungian analysts like June Singer (1976) have embraced the concept. Singer sees androgyny as implicit in Jung's notion of the self and says that, since each of us by nature is both masculine and feminine, we are inherently androgynous. Therefore, we do not need to become androgynous. We merely need to recognize our true nature.

She notes that in a marriage it is practical and necessary for some tasks and responsibilities to be divided between the husband and the wife. Of course, this division will vary according to the work undertaken by either partner. She sees a greater danger arising if the practical role division is carried over into the inner being of either partner. This would be the case if one partner assumes the role of doing all the thinking and making all the decisions, while the other partner lives out the tender emotional side of the marriage. If this split is carried to the extreme, we would have two inflexible partners coexisting but incapable of deep communication. If both partners are aware of their androgynous nature, neither one is seeking a partner to fill the empty space within — i.e., a partner to supply the masculine or feminine quality lacking in his or her own personality. Instead, each partner will be seeking someone to provide a masculine complement to the inner feminine side and a feminine complement to the inner masculine side.

Madonna Kolbenschlag (1979) advocates androgyny as a goal for people in general because she sees a need to modify traditional sex roles and to transcend the sex-stereotyped qualities traditionally associated with them. As more and more women proceed to work outside the home, they must be able to do so without being burdened with total responsibility for household maintenance and child-rearing. More tasks need to be shared by men and women. Kolbenschlag notes that women have been socialized to seek fulfillment in relationships with others, while men have been encouraged from childhood to seek fulfillment in achievement. Both men and women need to become more flexible and less bound by the traditional role division, so both husband and wife can seek satisfaction in both work and home.

Most of the writers who have advocated androgyny have been women, who are quite aware of the limitations imposed on women by traditional sex-role expectations. Of course, the goal they espouse does not serve the interests of women alone, for traditional sex-role expectations are also damaging to many men. A man may find it necessary to perform a menial and repetitive task at a low wage while expected to support a large family, and he may struggle to achieve status or prosperity in the face of seemingly insurmountable social, ethnic, and educational barriers. If he does achieve success in the eyes of society, he may do so at the expense of great stress, resulting in health problems that culminate in early death. If men are less vocal about the problems created by the sex-role expectations of our society, it may mean that they are less aware than the female critics or that they have been socialized to be tough and do the "manly" thing.

Many writers have talked about androgyny, and they have conceptualized it in a variety of ways. Some speak primarily in terms of a transcendence of traditional sex roles. Some speak of a balance of qualities that would serve to mitigate the problems presented by such traits as aggression, passivity, and emotional hypersensitivity. Some focus on behavioral change, while others focus more on modifications in the traditional schemata in terms of which we view ourselves and other people. Ellen Piel Cook (1985) has provided a useful survey of some of the theory and research relating to the concept of androgyny.

One of my concerns regarding the research in this area is that, while researchers usually recognize that they are dealing with a variety of sex-linked traits, they have usually treated masculinity and femininity as just two dimensions. In research they have commonly used two scales, each of which encompasses a conglomeration of traits. Then depending on how high or low an individual scores on the two scales, he or she is categorized as basically masculine, feminine, androgynous (high on both scales), or undifferentiated (low on both scales). Yet many combinations of so-called masculine and feminine traits are possible, and the value of any given combination may vary with the life situation and age status of the individual. Many traits that we consider masculine because they are seen more often in men (e.g, violence) or feminine because they are seen more often in women (e.g., extreme passivity) seem inherently undesirable. Perhaps what is more important than androgyny, as defined by a combination of two high scores, is flexible access to many different modes of functioning. I grant, however, that an individual who displays that sort of flexibility is likely to be loosely describable as androgynous.

FLEXIBILITY AND BALANCE

I believe it is true, as many advocates of androgyny claim, that many of us enter adulthood as rather lopsided individuals. We have cultivated some of our potentials at the expense of others. For a time this unbalanced condition may serve us well. Perhaps it enables us to secure a desired partner and launch a career. Yet over time we may experience growing dissatisfaction, stress, and friction with other people, and we need to allow room for psychological potentials that we have neglected.

I recognize that often the developed potentials may be broadly characterized either as masculine or as feminine and that it would be beneficial to allow more expression for the "contrasexual" side, but this broad characterization oversimplifies the matter. We are dealing with a number of distinguishable modes of being, and they cannot all be clearly identified as either masculine or feminine. In the preceding chapters we have considered a number of modes of being. Each one can serve a useful purpose and may be worthy of cultivation. At the same time, an exclusive or excessive emphasis on any single mode may lead to neglect of some of the individual's needs and its expression may have deleterious effects on that individual or on others.

Nurturance and compassion are certainly desirable qualities. It is essential for infants to be nurtured, and all of us are likely to need nurturing at other times in our lives. It is easy to admire people who devote their lives to helping others, but I remember a priest I met years ago whose whole life seemed devoted to tuning in to the needs of others and helping them in any way he could. Many of us who knew him feared that he was too ready to neglect or sacrifice his own needs at the risk of shortening his own life of service. For him, nurturance seemed to have assumed the status of a sacred duty that took precedence over anything else.

We see this kind of dedication more often in a woman who channels all her energies into motherhood and becomes such a devoted care-taker for her children that she is unable to relax her hold on them as they become more capable to acting independently. Ultimately she cannot let them go when she should. She would serve her own needs better if she cultivated her autonomy, as well as other interests — perhaps in the arts or in causes outside the home.

Those who emphasize ascendance may also provide a valuable function, for a group may accomplish little unless someone assumes the role of leader. We often need the person with the greatest maturity, knowledge, or experience to assume control of a situation. The individual who feels compelled at all times to be the dominant member of a couple or group, however, may have a problem. Such a person may have an underlying feeling of inferiority and may fear that he or she (more often he?) will lose power and control and appear inadequate to himself (or herself) and others. The underlying fear may need to be addressed, but the cultivation of other talents and skills may help alleviate the problem.

Aggression may serve some of the same functions as ascendance, and depending on the form it takes, it may lead to the correction of an undesirable situation — a situation in which someone in need is being neglected or someone is the object of unfair discrimination. Aggression is more often displayed by men, particularly when it assumes a physical form, but some women are quite adept at verbal aggression. I believe physical violence almost always creates more problems than it solves, and I see no justification for repeated violence directed at spouses, children, or members of ethnic minorities. In most cases, men who go out of their way to commit violent acts against gay men do so in an effort to deny their own homoerotic urges and prove they have no such tendencies. It would be easy for me to suggest that men who are violence-prone need to cultivate other modes of being, but they are likely to maintain the same behavioral pattern unless they recognize that they have a serious problem and that they may need therapeutic help to overcome it.

Many people value their autonomy, feeling it enables them to take full charge of their own lives. There is a clear advantage to spending much time alone if we wish to do creative work, follow a spiritual path, rest, or quietly reflect on our lives or the events of the day. It is possible that some people manage to lead very rich solitary lives, but many individuals who carry autonomy to an extreme do so because they are afraid of being hurt in relationships or because they feel socially inept. The cultivation of interpersonal skills through involvement in activities that demand interaction with others is one possible remedy. On the other hand, there are people who feel a need for constant interaction because they fear aloneness. They may be able to cultivate autonomy by recognizing things they can best accomplish by working alone.

The value of ordered, rational thought should be obvious to all of us, because we all attended schools that demand a great deal of it. Yet we do not all rely on it to the same extent, nor do we all find it equally essential in our work or our everyday lives. Certainly there are people who would do well to employ it more often, but there are also people who appear to use it constantly. They resort to analysis and intellectualization in all situations because they are afraid to deal with feelings. Years ago when I was in encounter groups, I found that such people would find great relief when they managed to share their feelings with others and discovered that they suddenly had more freedom to experience joy and the warmth of human interaction. If a balance is needed for over-reliance on rational thought, any activity, social or aesthetic, that permits more focus on feelings or more expression of them is likely to help.

I spoke of intuitive wisdom as a desirable complement to ordered, rational thought. Most creative work requires a balance of these two modes of cognition. One can obtain valuable insights by paying attention to unexpected ideas, images, and premonitions. Yet there is a danger in constantly relying heavily on the inspiration of the moment and not subjecting the ideas that pop into one's head to analysis and reason. We see this in individuals who cling to a sudden negative impression of the actions and motives of a friend or acquaintance, never stopping to examine the relevant evidence, never realizing that they are actually projecting motives from the recesses of their own psyches. We see a more extreme example in an individual whose grandiose visions are combined with a lust for power and a gift for persuasion. The charismatic leader Rev. Jim Jones seems a good example. He formed a religious group called the "People's Temple" and founded the Jonestown community in the jungle of Northwest Guyana. In 1978 the community rehearsed a mass suicide, and by the end of that year 913 people of an estimated 1100 in that community were dead by either suicide or murder.

Aesthetic and spiritual pursuits can enrich one's life. They may provide a needed balance for a life centered around activities that afford little expression of tender feelings (focused perhaps on such modes as ascendance, aggression, and rationality). I hesitate to suggest that anyone could be too heavily involved in the aesthetic realm, but one could argue that a person who devotes every waking moment to the arts would benefit from the dose of "reality" that a bit of time spent doing something different might offer. I have already noted my bias against those whose piety takes the form of preaching dogma — especially when, in the service of their beliefs, they seek to subvert science education, see their views cast into laws, and advocate the slaughter of those who don't think the way they do. There are times when ideas of either an aesthetic or allegedly spiritual nature need to be subjected to critical analysis.

There is an obvious value in being able to express oneself effectively in speech and writing. Perhaps such a skill is over-emphasized by people who constantly seek to debate with their acquaintances so they can display their gift for glib speech and clever arguments. Many more of us avoid developing expressive talents for fear that we will not succeed.

In Chapter 9 I spoke of wildness and impulsivity. Robert Bly and Clarissa Pinkola Estés rightly stress the value of getting in touch with our wild side, that source of natural inclinations that underlies the many layers of cultural conditioning we acquired in the course of growing up and fitting into adult society. Most of us would benefit from getting in closer touch with the Wild Man or the Wild Woman within. On the other hand, there are some people who seem under-socialized and almost constantly act on the impulse of the moment. The action may be fun, but the consequences are often not. A balance might be provided by rational thought or by the discipline afforded by various forms of education and activity. But surely nobody with such a problem is going to read this book. You, dear reader, are certainly not such an individual.

The realm of love and sex is obviously at times a source of much emotional arousal for most of us. Often it becomes the focus of severe emotional problems. We tend, especially in our youth, to experience romantic love as something that happens to us, something over which we have no control. We may experience great inner turmoil, depending on whether our beloved is actually aware of how we feel and whether he or she reciprocates. The situation is particularly difficult if in falling in love we have projected an overly idealized image onto the other person. Then, our expectations will keep clashing with reality until we are forced to face reality and accept it. There is no easy remedy for a romantic obsession, but it helps to have other meaningful pursuits to which one can devote time. I believe most adults over the years become more aware of choosing to love and more capable of loving the other person for who he or she actually is.

Then it is easier to "stand" in love, rather than just "fall" in love. Yet the moment of unexpected delight when we see a certain face, a certain smile, or a certain gesture will always hold a kind of magic and charm that we may not want to give up altogether.

Love and sex are often intertwined, but many of our attitudes regarding sex and the accompanying behavior have little to do with love. Many people are reared to regard sex as dirty or sinful. Sexual abuse in the growing years may leave an impression that is hard to erase. Various experiences may lead either a boy or a girl to feel physically lacking in the sexual realm. Thus for many people, sexual expression becomes something to avoid. At the other extreme are people compulsively driven to engage in sex with many partners. A woman may have sex with many men to prove to herself that she is desirable and that she is loved. A man may pursue sex with many women in order to quell any doubts regarding his virility or to deny underlying homosexual impulses. An individual may need therapeutic help to overcome either a fearful avoidance of sexual expression or a compulsive over-concentration on it.

We all pass through childhood before we become adults, and the children we once were remain a part of us. The inner child can be a source of spontaneity and freshness, enabling us to re-experience beauty and awe in the face of a world that has become too familiar. Too many people lose touch with the inner child. But there is a big difference between being in touch with the child and being stuck in the role of a child. An individual with normal physical and mental capacities who continues to display the helplessness and neediness of a child is failing to take responsibility for independent action. Such a person needs to be able to utilize whatever modes of being are called for in his or her current life situation — perhaps autonomy, nurturance, or rational thought.

The talent of the psychopomp or the shaman may seem beyond the reach of most of us. Yet we all pass through various states of consciousness, and there are people who develop an ability to enter sleep with full consciousness. Some Tibetan Buddhists cultivate this ability through the practice of meditation. Many of us have managed to cultivate lucid dreaming, being aware while dreaming that we are dreaming. It is also possible to develop an ability to tune in to the inner worlds of other people and enter those worlds. Many actors become adept at doing this. The ability to cross psychological boundaries can enrich one's life, but the inability to control the crossing can be a serious problem. I suggested this with respect to Odysseus. In reading about Odysseus, who claims to be Nobody, I am reminded of schizophrenics who can pretend to be "crazy" but who can also act normal for the moment if the situation demands it. Despite this evident ability to shift, however, inside they have lost a clear feel for the dividing line between reality and fantasy.

Many people who have experimented with substances like marijuana have had the experience, while fully awake but intoxicated, of climbing back to ordinary reality, feeling they have arrived, then realizing they must climb another flight of stairs. The difficulty is much greater when one uses a stronger psychedelic drug like LSD. Many people who took repeated lysergic acid "trips" in the 1960's and 1970's were unable to return fully to ordinary reality after the drug had left their bodies, and they had to battle psychosis for a time after they had ceased using the drug.

In this section I have tried to offer a few more thoughts about each mode of being, noting conditions that may prevent or limit our use of a given mode as well as conditions that may lead us to rely on it excessively. We all grow up with a few fears that block us from exploring and developing all of our potentials. These can include a fear of being effeminate or sissified, a fear of asserting ourselves because we feel physically or intellectually inadequate, a fear of being seen as too boyish and not lady-like, a fear of homosexuality, a fear of being rejected or hurt if other people find out that we do have homosexual urges, a fear of being alone, a fear of losing our identity if we are constantly surrounded by other people, a fear of love, and a fear of sex.

To the extent that many of these fears are bound up with indoctrination in traditional gender roles, I applaud the advocates of androgyny for pointing to a liberating path. A woman who feels she must be a Stepford wife and a man who feels bound to play the role of a tough macho man are both unable to experience the richness of life that greater flexibility would afford. Yet gender roles are only a part of the total picture. There are many modes of being to consider, and it does not help to try classifying them all as either masculine or feminine. They are all part of us, and you have had a chance to think about them. There may not be enough time in one life to cultivate them all to the extent we might wish. Which ones have you emphasized in your own development? Which ones would be a valuable addition to your life if you gave them more attention?

Possible Modes of Fulfillment

The main condition I have stressed as a desirable goal is flexible access to various modes of being. This does not mean that they must all be cultivated, nor that we should avoid concentrating heavily on any one. Obviously an artist, a composer, or a poet must still emphasize the aesthetic realm. A courtroom lawyer must cultivate expressiveness, rational thought, and perhaps some form of aggression.

There is no general agreement with respect to a type of person or a mode of adjustment that represents the one goal to which we should all aspire. At various times, people have referred to the goal as maturity, normality, positive mental health, and self-actualization, but none of these labels represents a very clearly definable concept. Psychologists and theorists in related disciplines have noted many traits that they consider ingredients of the ideal, or optimal, personality. Some years ago, I did a large-scale analysis of many such variables, and my analysis yielded a number of independent factors (Coan, 1974). The variables did not all go together. You could be very high on some of them while being low on others. It was clear that various personality styles might be deemed ideal, depending on who was judging the matter.

Following that research, I surveyed views of the ideal condition running through both Eastern and Western traditions, as well as views evident in various disciplines in recent times (Coan, 1977). I concluded that the many views regarding the optimal mode of being could be sorted into five major modes of fulfillment, which I named as follows:

(1) efficiency
(2) creativity
(3) inner harmony
(4) relatedness
(5) transcendence

The mode of efficiency encompasses many models, both ancient and modern, involving intellectual, physical, or social competence. We see it in various forms in the ideals of classical Greece, the ideal of medieval knighthood, the lauded versatile genius of the Renaissance, and various idolized images of modern times. Depending on the specific form of efficiency that is emphasized in a given ideal image, physical and work skills may be stressed. For some fields of endeavor, either ascendance or autonomy may be demanded. If the focus is on intellectual competence, the cultivation of ordered, rational thought is probably essential.

The mode of creativity overlaps with efficiency to the extent that both commonly imply something in the way of production. Creativity, in addition, entails originality. It involves the discovery or creation of novel form. To achieve this, an individual must be able to experience something in a fresh way, at least within the realm in which he or she does creative work. The most highly creative people are likely to experience much of life and the world in fresh ways. They must be able to free themselves from the prevailing views and ways of perceiving in their society and in their chosen discipline or profession. They must be open to the novel and unexpected ideas and images that loom into consciousness. As I indicated in an earlier chapter, this requires what I have called intuitive wisdom, and if inspiration is to be followed by the successful production of new form, intuitive wisdom must be accompanied by ordered, rational thought. Some degree of autonomy is also necessary if the creative person is to function very independently. Depending on the field in which the creative individual operates, various other modes of being may be involved — perhaps the openness of the child, nurturance, aesthetic pursuits, expressiveness, wildness, or sensuality.

Inner harmony is difficult to assess. The individual who is well adjusted to his or her niche in society, conforms to prevailing norms, assumes a traditional gender role, and possesses high self-esteem may appear to manifest considerable inner harmony. That harmony, however, may be limited to a surface layer of the personality. It rests on a blissful unawareness of potentials and tendencies that escape conscious recognition. A more inclusive inner harmony would require openness to the total realm of experience. Such an openness is implied in Jung's notion of the realization that the self is the center of the psyche. This would involve awareness of primitive urges and contrasexual tendencies that most people are aware of only in projection onto other people. It would also involve allowing a sensing and expression of forces within the psyche that lie beyond the reach of consciousness. It would involve a recognition of contrasting or competing elements, urges, and tendencies within one's total being and seeing them as complements that balance one another, rather than acknowledging only some and denying or repressing others. With respect to the modes of being I have discussed in the last few chapters, inner harmony might entail a cultivation and balance of modes that seem often to lead in opposite direction — e.g. nurturance and autonomy. It might also entail the use of modes that imply various forms of openness — e.g., intuitive wisdom, aesthetic and imaginal pursuits, piety, wildness, sensuality, and the inner child.

Relatedness is obviously concerned with the interpersonal realm. One form in which it is manifested is the intimacy possible with another individual whom one understands and values as he or she is. This implies a genuine love for the actual person, not the person as the bearer of an unreal image that one has projected. On a more general level, relatedness implies an acceptance of, and caring for, people in general. It may be extended even more broadly to encompass all living creatures, wild and domestic. Of the modes of being I have listed, the one most central to relatedness is nurturance and compassion.

What I am calling transcendence overlaps with relatedness. It involves experiences that have been described in terms of a relationship to God, to the divine, to the whole of nature, and to the ultimate ground of being. Some religious experiences may be characterized as transcendence, but transcendence need not entail any concept of God or even an acceptance of any traditional religious view. Much of religious experience is better characterized in terms of relatedness than in terms of transcendence. Martin Buber (1937), a major religious writer and philosopher, speaks of God as the "eternal Thou" and speaks of his relationship to God in terms of an "I-Thou" experience. This amounts to relatedness between the individual and a divine other.

We move from relatedness to transcendence to the extent that a sense of individual separateness vanishes and there is no dividing line between "I" and other. Much of what is called pure spirituality and mysticism involves transcendence. In major Eastern traditions, individual identity is considered an illusion that vanishes when one achieves enlightenment (or comes to realize that one is already enlightened).

Viewed historically, a religion usually begins with a transcendent spiritual experience. By its very nature, that experience is ineffable. There is no way to describe it in ordinary language, for our words are designed to deal with distinctions and differences. But in an effort to communicate the experience symbolically, we create myths, and these lead in turn to theological doctrines. For many people, religion then becomes an intellectual exercise or a body of dogma, far removed from the realm of transcendence.

Transcendence may be a part of many experiences we do not usually regard as religious, experiences we might characterize as "peak" experiences or ecstasy. We may have such an experience in the enjoyment of the arts, in a mystical union with nature, or in an intimate encounter in which for the moment we experience a merging with a partner and cease to feel like two separate people. Several of the modes of being I have discussed may involve experiences that have some connection to transcendence, notably piety, aesthetic and imaginal pursuits, and intuitive wisdom. Nurturance and compassion are obviously concerned with relatedness, but their expression sometimes involves transcendence as well.

I have discussed these five modes of fulfillment to underscore the fact that there is no universal agreement regarding the ideal goal of personal development and to call attention to some of the major variations in views on the subject. Emphases vary cross-culturally. The individualistic traditions of the West tend to value efficiency and creativity. In the East, we find more stress on transcendence. Relatedness and inner harmony have been emphasized to some extent in both East and West and perhaps more so recently than in the past.

In Chapter 10, I noted the views of Erik Erikson and Carl Jung regarding life stages. Erikson's concept of generativity suggests an emphasis on relatedness as a life goal. Jung's treatment of the individuation process suggests more emphasis on inner harmony and possibly transcendence. In Hindu texts, most notably the Upanishads, it is even clearer that transcendence is regarded as the proper focus of the later years. Having passed through a stage as a student and a stage as a householder, according to those texts, the individual then enters a stage of retirement in the forest, followed by a fourth stage of renunciation of the world.

An over-emphasis on one mode in a given cultural setting may call for a fresh emphasis on another mode as people sense a need for balance. The modes of relatedness, transcendence, and inner harmony, for example, may offer a desired balance to the more individualistic modes of efficiency and creativity. Within any one society values vary from one group to another. Your own values may shift over your lifetime, so that the mode you favor in your later years differs from the one you favored in your youth.

I would not advocate any one mode of fulfillment over the others, and I would not suggest that you must cultivate a particular mode of being and forsake others. Only you can decide what is most suitable in your own life. On the other hand, perhaps each of us goes through periods in which a too exclusive emphasis on one mode of being results in great dissatisfaction. For such times, I would suggest that a consideration of other available modes may reveal a needed path. In general, I believe that cultivating a variety of different potentials, different modes of being, can lead to a reduction of distress and greater joy in living.

References

Bakan, David. *The Duality of Human Existence.* Chicago: Rand McNally, 1966.

Baron-Cohen, Simon. *The Essential Difference: The Truth about the Male and Female Brain.* NY: Basic Books, 2003.

Bem, Sandra L. The Measurement of Psychological Androgyny. *Journal of Consulting and Clinical Psychology,* 1974, Vol. 42 (No. 2), 155-162.

Bly, Robert. *Iron John: A Book about Men.* Reading, Mass.: Addison-Wesley, 1990.

Bolen, Jean Shinoda. *Goddesses in Everywoman: A New Psychology of Women.* San Francisco: Harper & Row, 1984.

Bolen, Jean Shinoda. *Gods in Everyman: A New Psychology of Men's Lives and Loves.* San Francisco: Harper & Row, 1989.

Buber, Martin. *I and Thou.* Edinburgh: Clark, 1937.

Cattell, Raymond B. and Richard W. Coan. Objective-Test Assessment of the Primary Personality Dimensions in Middle Childhood. *British Journal of Psychology,* 1959, Vol. 50, 245-252.

Coan, Richard W. *The Optimal Personality: An Empirical and Theoretical Analysis.* London, England: Routledge & Kegan Paul, 1974. (Co-published by Columbia University Press, New York)

Coan, Richard W. *Hero, Artist, Sage, or Saint? A Survey of Views on What Is Variously Called Mental Health, Normality, Maturity, Self-Actualization, and Human Fulfillment.* NY: Columbia University Press, 1977.

Coan, Richard W. Dimensions of Masculinity and Femininity: A Self-Report Inventory. *Journal of Personality Assessment,* 1989, Vol. 53 (No. 4), 816-826.

Coan, Richard W. *Shaul of Tarsos: The Man Who Came to Be Known as Saint Paul.* Bloomington, Indiana: AuthorHouse, 2004.

Davis-Kimball, Jeannine. *Warrior Women: An Archeologist's Search for History's Hidden Heroines.* NY: Warner Books, 2002.

Downing, Christine. *The Goddess: Mythological Images of the Feminine.* NY: Crossroad, 1981.

Downing, Christine. *Gods in Our Midst: Mythological Images of the Masculine: A Woman's View.* NY: Crossroad, 1993.

Erikson, Erik H. *Childhood and Society.* NY: W. W. Norton, 1963.

Erikson, Erik H. Inner and Outer Space: Reflections on Womanhood. *Daedalus,* Spring, 1964, 582-606.

Estés, Clarissa Pinkola. *Women Who Run with the Wolves: Myths and Stories of the Wild Woman Archetype.* NY: Ballantine, 1992.

Fehr, Ruth E. Stimulus Determinants of Color Harmony. Thesis submitted for the MA degree at the University of Arizona, 1963.

Friedan, Betty. *The Feminine Mystique.* NY: Norton, 1963.

Gray, John. *Men Are from Mars, Women Are from Venus.* NY: Harper Collins, 1992.

Kinsey, Alfred, Wardell B. Pomeroy, and Clyde E. Martin. *Sexual Behavior in the Human Male.* Philadelphia: Saunders, 1948.

Kinsey, Alfred, Wardell B. Pomeroy, Clyde E. Martin, and Paul H. Gebhard. *Sexual Behavior in the Human Female.* Philadelphia: Saunders, 1953.

Kolbenschlag, Madonna. *Kiss Sleeping Beauty Good-Bye.* San Francisco: Harper, 1979.

Lips, Hilary M. *Sex and Gender: An Introduction.* NY: McGraw-Hill, 2005.

Maccoby, Eleanor Emmons, and Carol Nagy Jacklin. *The Psychology of Sex Differences.* Stanford, California: Stanford University Press, 1974.

Malinowski, Bronislaw. *Sex and Repression in Savage Society.* London: Routledge & Kegan Paul, 1927.

Mead, Margaret. *Sex and Temperament in Three Primitive Societies.* NY: Morrow, 1935.

Moore, Robert, and Douglas Gillette. *King, Warrior, Magician, Lover: Rediscovering the Archetypes of the Mature Masculine.* San Francisco: Harper, 1990.

Mosse, George L. *The Image of Man: The Creation of Modern Masculinity.* NY: Oxford University Press, 1996.

Neumann, Erich. *The Great Mother: An Analysis of the Archetype.* Princeton, New Jersey: Princeton University Press, 1963.

Stephens, James. *The Crock of Gold.* London: Macmillan, 1912.

Stoller, Robert J. *Sex and Gender: On the Development of Masculinity and Femininity.* NY: Science House, 1968.

Terman, Lewis, and Catherine Cox Miles. *Sex and Personality: Studies in Masculinity and Femininity.* NY: McGraw-Hill, 1936.

Weber, Max. *The Protestant Ethic and the Spirit of Capitalism.* NY: Scribner, 1958. (originally published in German in 1905)

About the Author

Richard W. Coan is a Professor Emeritus of Psychology and resides in Tucson, Arizona.

As a child he began composing music and writing stories and poems. Though drawn to the arts, he was also intrigued by the puzzles of human consciousness, the sense of personal identity, and the varied idiosyncrasies of the people he knew. His search for understanding led to a career in psychology. At both the University of California and the University of Southern California, he pursued graduate studies leading to a doctorate in clinical psychology in 1955.

As a professor at the University of Arizona he specialized in personality theory and measurement. His published articles and books cover such topics as the evolution of consciousness, the optimal personality, masculinity/femininity, child personality, patterns of orientation among psychologists, and trends in psychological theory. His interests include Jungian theory, the psychology of religion, Eastern thought, and archetypal symbolism in myths, tales, and dreams. In this book, he has combined his interest in major mythic figures with his lifelong effort to understand the basic variation in human personality and the major life choices that people make.

Throughout his academic career and in retirement, he has continued writing poetry and composing music. In recent years he has published three novels: (1) A Princess for Larkin, (2) Shaul of Tarsos: The Man Who Came to Be Known as Saint Paul, and (3) Horatio: The Loyal Friend of Prince Hamlet.

Index

tricksters 60, 107, 108, 109, 110, 111,
     114
Tristram 71
Turner's syndrome 37, 38
Tyr 78

U

Uranus 74

V

Valentino, Rudolph 103
Vedas 75
verbal ability 16
Vesta 93
Vishnu 75

W

Wallenberg, Raoul 73
warrior archetype 116
Weber, Max 49, 57
wildness 99-101, 111, 114, 133
Wild Man, the 114
Wild Woman, the 118, 133
Williams, Robin 99, 100
Winfrey, Oprah 98
witches 51, 118
wolf-woman self, the 118

Z

Zeus 68, 69, 74, 75, 77, 79, 80, 83,
     84, 87, 91, 98, 99, 102, 104,
     108, 117

CPSIA information can be obtained at www.ICGtesting.com
Printed in the USA
LVOW05s1811180913

353047LV00001B/146/P

9 781438 921730